Quantum Mechanics

David B. Beard

With a Preface by **Dian Curran**

Dover Publications, Inc.
Mineola, New York

TO { HANS A. BETHE
RICHARD P. FEYNMAN
J. J. GERALD McCUE

PREFACE TO THE DOVER EDITION

Dr. David Breed Beard was a child of the Great Depression. After serving his country in the Navy in World War II, he finished his education. He attended Hamilton University with the aid of scholarships and received his Ph.D. in nuclear physics from Cornell University under the guidance of Hans Bethe. He later shifted his focus to theoretical space physics.

David Beard accepted a professorship at the University of California, Davis, and later moved to the University of Kansas to serve as Chairman of the Department of Physics and Astronomy and became a University Professor before retiring in 1987. Many of his sabbaticals were spent collaborating with researchers at Imperial College in London. Among his most notable accomplishments were his derivation of a model of planetary magnetospheres and his study of comets.

In 1963, David Beard published this book, *Quantum Mechanics*. It was an important new textbook in the field of physics at the time, featuring the Feynman path integral to introduce the wave function and emphasizing the close relationship of quantum theory and physical optics.

David Beard also had the satisfaction of having one of his children follow him into his professional field and later work with him—his daughter Dian. Dr. Dian Curran worked with her father on modeling the interaction of the solar wind with mercury. Although mercury has no intrinsic magnetic field, its high iron content does interact with the solar wind, creating a weak magnetosphere surrounding the planet.

Outside of work, David Beard loved to go on country hikes which his daughter Dian has many fond memories of, and many others who knew David remember him for his story-telling. Other than Dian Curran, David had three other children, Lawrence Bennett Beard, Jonathan Breckenridge Beard and Valerie Curran Beard and a grandchild from Jonathan, Christopher Patrick Beard.

David Breed Beard is gone now, but his good work, of which this book is a part, remains.

Dr. Dian Curran, September 2013

PREFACE

THIS BOOK WAS WRITTEN with the aim of providing an alternative to the postulatory approach currently used in many introductory quantum mechanics courses. Thus the first chapters are designed to familiarize the reader with the mathematical treatment and particular properties of ordinary wave motion which apply also to particle motion and have crucial importance to the subsequent development of the subject. The close relation of quantum theory to physical optics is stressed, and by use of Feynman's derivation of quantum mechanics, the wave theory for particles is made to appear as inevitable and necessary as Huygens' wave theory for light.

Quantum mechanics is as complete a closed system as Euclidean geometry, and as such should be given the same integrated, balanced development in a separate course devoted to it. However, although the postulational, geometry-inspired approach is traditional in introducing quantum mechanics (and superbly executed examples of this method may be found in several currently available texts), it is not wholly satisfactory. The subject is introduced by non-self-evident postulates detailing the properties of something called a wave function; then it is said that the wave function obeys a second-order differential equation known as the Schroedinger wave equation; finally, the theory is applied to some examples, and away the subject goes! Almost any student, when asked why he accepts all this, replies, "Because it works," which is ultimately the only reason for accepting it; however, he still has far to go before arriving at a complete idea of the subject and an understanding of why the theory is what it is.

On the other hand, the student has seen wave motion, understands it, and knows that the only reason for accepting the wave theory of light in preference to the ray theory is that diffraction and interference of light can be explained by the wave theory. He has worked with phase differences of permissible light paths, for example when using Fraunhofer diffraction theory to analyze a diffraction grating. Therefore, once he is shown in detail how a ray theory of particles breaks down for electrons transmitted through a crystal, he should be quite prepared to accept a simple Fraunhofer diffraction theory for particles identical to that for light. This intuitively appealing argument was first presented by Feynman in 1948. While at least as rigorous as the initial arguments by Schroedinger and Heisenberg in 1926, the argument focuses on the wave function directly, and thus the student who has previously worked simple problems in physical optics should have little difficulty realizing what the wave function is supposed to represent, what its properties are, and how it behaves in a homogeneous or inhomogeneous medium (i.e., how it depends on the light or particle path in a medium of constant or changing index of refaction or particle potential energy).

Following this introduction, the text stresses the physical consequences of a wave theory of material particles by examining the extreme, non-classical characteristics of abrupt changes in potential energy as a function of space. Classical mechanics is later shown to arise as an approximation valid when lengths, momentum, energy, or time are not specified with a precision greater than that permitted by the Uncertainty Principle. This semiclassical approximation, standard examples, and other approximation methods are discussed in order to develop the student's confidence and his ability to apply the theory to diverse physical problems. The subjects treated in the introductory chapters, for instance solid state theory and the theory of the deuteron, were chosen for their mathematical simplicity. The physical applications stressed in later chapters are introduced to prepare the student in topics he is likely to encounter early in his further study of atomic and nuclear physics.

The first eleven chapters of the book—through the treatment of chemical bonding—constitute sufficient subject matter for an undergraduate, one-semester course. The last two chapters can then be assigned as optional, although they contain much material vital to any subsequent graduate study in physics.

Many people have contributed toward whatever excellences the book may have. My considerable debt to lectures by, and conversations with, Professors Hans A. Bethe, Richard P. Feynman, and Linus Pauling, and to the excellent texts by Peter Bergmann, David Bohm, and Linus Pauling and E. Bright Wilson, will be evident to anyone familiar with these persons' understanding of the subject. Dr. William A. Newcomb assisted me invaluably in finding errors and obsurely presented material. My editor Mrs. Cynthia Easton painstakingly edited the entire manuscript. My colleague Professor William W. True read and commented most helpfully on every chapter. The students in my classes, especially Mr. Willard Sperry, spurred me to clearer presentation of many initially obscure parts of the texts. I am continually indebted to Professors J. J. Gerald McCue and Richard O. Sutherland, who first stimulated my interest in quantum mechanics. My further thanks are due to the publishers' reviewers of the manuscript for their valuable suggestions, to my wife for her patient assistance, and to Mrs. Natalie Fulk, Mrs. Viona Hague, and Mrs. Iva Parelius for exceptional clerical aid.

David B. Beard

CONTENTS

4   THE FEYNMAN PATH INTEGRALS AND THE SCHROEDINGER EQUATION

5   USE OF THE WAVE FUNCTION TO OBTAIN PHYSICAL PROPERTIES OF A SYSTEM

6   THE CONSEQUENCES OF ABRUPT CHANGES IN POTENTIAL ENERGY

11 CHEMICAL RESONANCE THEORY

12 SPIN, SYMMETRY, PARITY, AND VECTOR ADDITION

13 ELASTIC SCATTERING THEORY

Quantum Mechanics

BLACK BODY RADIATION

HISTORICALLY, THE QUANTUM THEORY first arose from studies of radiation, and it is useful to introduce the subject along historical lines in which relevant aspects of electromagnetic waves may be reviewed. Quantum mechanics has, with good cause, also been called wave mechanics since it emphasizes and treats wave-like behavior of particles. Hence, it is appropriate to begin our study by an examination of familiar classical radiation theory. By this means we will be able to investigate relevant consequences of wave motion and develop for later advantage the mathematical tools frequently used in classical treatments of electromagnetic waves.

1.1 DERIVATION OF THE ELECTROMAGNETIC WAVES IN A CAVITY

Suppose a light wave to be incident on a small hole in the side of a box which is otherwise completely light-tight so that radiation enters or leaves the box only through the small hole. Such a box is called a black body since the hole completely absorbs all light incident on it. The incident light wave will be lost inside the box after repeated reflections from the walls and will come into thermal equilibrium with the walls. Thus a light wave escaping outside the box through the hole will have been in thermal equilibrium with the interior of the box. The radiation in equilibrium within the box consists of electromagnetic waves or, in the parlance of classical radiation theory, oscillations of the ether, the vibrating medium within the box by which the waves are propagated and sustained. *In order to estimate the radiation to be expected from the pinhole, it is necessary to calculate first what are the possible radiation frequencies and what is the energy of each separate electromagnetic wave.* The energy or light intensity emitted from the opening at a particular frequency is proportional to the total energy of all the waves having that frequency contained within the box. Hence, the intensity of the electromagnetic radiation within the box must first be estimated by means of Maxwell's equations for the electric and magnetic field vectors of the waves in the interior, for the conditions of free space within the box. Expressed in Gaussian units Maxwell's equations are

$$\nabla \times \mathbf{H} = \frac{1}{c} \frac{\partial \mathbf{E}}{\partial t} \tag{1.1}$$

$$\nabla \times \mathbf{E} = -\frac{1}{c} \frac{\partial \mathbf{H}}{\partial t} \tag{1.2}$$

$$\nabla \cdot \mathbf{H} = 0 \tag{1.3}$$

$$\nabla \cdot \mathbf{E} = 0 \tag{1.4}$$

Taking the curl of both sides of Eq. (1.2), interchanging the order of the space and time derivatives, and substituting Eqs. (1.1, 1.2) in the result, by the rules of vector analysis we obtain three equations for the three components of the electric field vector, summarized in

$$\nabla^2 \mathbf{E} = +\frac{1}{c^2} \frac{\partial^2 \mathbf{E}}{\partial t^2} \tag{1.5}$$

This is a very familiar differential equation in physics, occurring for such diverse motions as a vibrating string, pressure variations in an organ pipe, flow of electricity in a cable, or the waves on a drum head. Its solution describes wave motion, and it is called the wave equation.

In Cartesian coordinates Eq. (1.5) has the particular form or representation

$$\frac{\partial^2 E}{\partial x^2} + \frac{\partial^2 E}{\partial y^2} + \frac{\partial^2 E}{\partial z^2} = \frac{1}{c^2}\frac{\partial^2 E}{\partial t^2} \qquad (1.5a)$$

Equation (1.5a) may be solved for one component E_x by assuming E_x to be written as a product of functions of one variable each. (This is known as assuming the variables separable.) Thus, let $E_x = X(x)Y(y)Z(z)T(t)$. This procedure will be justified if a solution is obtained; not all partial differential equations have solutions which can be obtained in this way. Dividing Eq. (1.5a) for the x component of the field by E_x, one finds in Cartesian coordinates

$$\frac{1}{X}\frac{\partial^2 X}{\partial x^2} + \frac{1}{Y}\frac{\partial^2 Y}{\partial y^2} + \frac{1}{Z}\frac{\partial^2 Z}{\partial z^2} = \frac{1}{Tc^2}\frac{\partial^2 T}{\partial t^2} \qquad (1.6)$$

The left-hand side is completely independent of the time, while the right-hand side is independent of space. Therefore, both sides of the equation are equal to a constant, independent of both time and space. Let the constant be written as $-\omega^2/c^2$. Then the time-dependent part of (1.6) becomes

$$\frac{1}{Tc^2}\frac{\partial^2 T}{\partial t^2} = -\frac{\omega^2}{c^2} \qquad (1.7)$$

whose solution is

$$T = D_1 e^{i\omega t} + D_2 e^{-i\omega t} \qquad (1.8)$$

where D_1 and D_2 are arbitrary constants for the time-dependent part of E_x. Similarly,

$$\frac{1}{X}\frac{\partial^2 X}{\partial x^2}$$

is equal to a constant, $-k'^2$, and thus the solution to the space-dependent part of (1.6) for E_x is

$$X = A_{1k'x}\ \sin k'x +\ A_{2k'x}\cos k'x$$
$$Y = B_{1l'x}\ \sin l'y +\ B_{2l'x}\cos l'y \qquad (1.9)$$
$$Z = C_{1m'x}\sin m'z + C_{2m'x}\cos m'z$$

where l' and m' are additional constants introduced for the Y, Z terms. Identical solutions may be obtained for E_y and E_z.

Since the tangential components of $\mathbf{E}$ must be continuous at the walls, $\mathbf{E}$ at the sides of the box cannot be arbitrary but is fixed by the physical properties of the wall material. In a box with infinitely conducting walls, for example, the tangential components of $\mathbf{E}$ at the walls must be zero (that is, E_x must be zero in the four walls of the box perpendicular to the y or z axes). Hence, in a rectangular infinitely conducting box of dimensions a, b and d, $Y(0) = 0$ and therefore $B_{2l'x} = 0$, and $Y(b) = 0$ and therefore $\sin l'b = 0$ which means that $l'b = \pi l$ where l is an integer. Thus, two of Eqs. (1.9) become

$$Y = B_{1l'x} \sin \frac{\pi l y}{b}; \qquad Z = C_{1m'x} \sin \frac{\pi m z}{d}$$

Similar results are obtained for the other two components or $\mathbf{E}$. Let $B_{1l'x}$, $C_{1m'x}$, $A_{1k''y}$, $C_{1m''y}$, $A_{1k'''z}$, and $B_{1l'''z}$ equal unity, which may be done with no loss of generality by readjusting the values of the remaining undetermined constants. Then the total solution for $\mathbf{E}$ may be written as

$$\mathbf{E} = (A_{1k'x} \sin k'x + A_{2k'x} \cos k'x) \sin l'y \sin m'z\, \hat{\imath}$$
$$+ (B_{1l''y} \sin l''y + B_{2l''y} \cos l''y) \sin k''x \sin m''z\, \hat{\jmath}$$
$$+ (C_{1m'''z} \sin m'''z + C_{2m'''z} \cos m'''z) \sin k'''x \sin l'''y\, \hat{\mathbf{k}}$$

The total solution for $\mathbf{E}$ must also satisfy Eq. (1.4), from which we obtain (after dividing out the time-dependent part)

$$A_{1k'x}k' \cos k'x \sin l'y \sin m'z + B_{1l''y}l'' \sin k''x \cos l''y \sin m''z$$
$$+ C_{1m'''x}m''' \sin k'''x \sin l'''y \cos m'''z$$
$$- A_{2k'x}k' \sin k'x \sin l'y \sin m'z$$
$$- B_{2l''y}\, l'' \sin k''x \sin l''y \sin m''z$$
$$- C_{2m'''z}m''' \sin k'''x \sin l'''y \sin m'''z = 0$$

Since this equation must hold for any values of x, y, and z, the coefficients of each term must each individually be equal to zero, and therefore only the

$A_{2k'xk'}$, $B_{2l''yl''}$, and $C_{2m'''zm'''}$ can be nonzero (although their sum must be zero) provided that $k''' = k'' = k'$, $l''' = l'' = l'$, and $m''' = m'' = m'$. Thus the spatial dependence of **E** is given by

$$E_x = A_{klm} \cos \frac{k\pi x}{a} \sin \frac{l\pi y}{b} \sin \frac{m\pi z}{d}$$

$$E_y = B_{klm} \sin \frac{k\pi x}{a} \cos \frac{l\pi y}{b} \sin \frac{m\pi z}{d} \qquad (1.10)$$

$$E_z = C_{klm} \sin \frac{k\pi x}{a} \sin \frac{l\pi y}{b} \cos \frac{m\pi z}{d}$$

where

$$k = ak'/\pi, \quad l = bl'/\pi, \quad m = dm'/\pi,$$
$$A_{klm} = A_{2k'x}B_{1l'x}C_{1m'x}, \quad B_{klm} = A_{1k''y}B_{2l''y}C_{1m''y}$$

and

$$C_{klm} = A_{1k'''z}B_{1l'''z}C_{2m'''z}$$

Equations (1.8, 1.10) together describe a sinusoidal wave of angular frequency ω ($\omega = 2\pi\nu$ where ν is the number of cycles per second and ω the number of radians per second), and velocity c. Equations (1.10) alone would describe a standing wave such as the particular example illustrated in Figure 1.1 for $C_{klm} = C_{410}$,

$$E_z = C_{410} \sin \frac{4\pi x}{a} \sin \frac{\pi y}{b}$$

By substituting Eqs. (1.7) and (1.10) into Eq. (1.5), one finds that

$$\left(\frac{k\pi}{a}\right)^2 + \left(\frac{l\pi}{b}\right)^2 + \left(\frac{m\pi}{d}\right)^2 = \frac{\omega^2}{c^2} \qquad (1.11)$$

For a particular frequency ω, all integral values of k, l, and m are possible (4, 1, 0) which satisfy Eq. (1.11). For a cube, for example, besides the mode (4, 1, 0) for E_z illustrated in Figure 1.1, there are four other modes for E_z, (1, 4, 0), (3, 2, 2), (2, 3, 2), and (2, 2, 3) with frequency $\omega = \sqrt{17}(\pi/a)c$. (The modes (4, 0, 1), (1, 0, 4), (0, 4, 1), and (0, 1, 4) are not included since $E_z \equiv 0$ for these modes.)

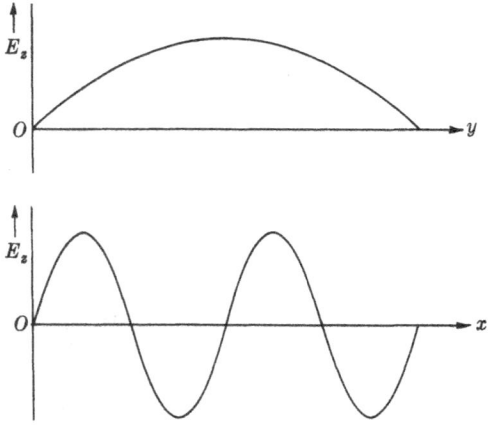

Figure 1.1 *Graphs of E_z as function of x and y for the mode $k = 4$,*
$l = 1$, and $m = 0$.

1.2 NUMBER OF MODES HAVING THE SAME FREQUENCY

In order to obtain the total energy of radiation of a particular frequency
ω contained within and emitted from the box, it is necessary to find the
number of different modes of oscillation all having the same frequency ω.
The number of modes having a frequency equal to or less than ω is found by
summing all integers l, k, and m which make the left-hand side of Eq. (1.11)
less than or equal to the right-hand side. For large values of ω such that the
wavelength $c/2\pi\omega$ is orders of magnitude smaller than the box dimensions
a, b, and d, the summation may be replaced by a triple integral over k, l,
and m, that is, a volume integral over a k, l, m coordinate space. Equation
(1.11) is the equation for an ellipsoid in which k, l, and m are the coordinates
and the semi-axes are $(a\omega/\pi c)$, $(b\omega/\pi c)$, and $(d\omega/\pi c)$ respectively (Fig. 1.2).
The number of modes having a frequency equal to or less than ω is the num-
ber of different combinations of positive integers k, l, and m such that the
left-hand side of Eq. (1.11) is less than or equal to ω^2/c^2. This number is
given by the volume of the positive octant of the ellipsoid,

$$(\pi/6)\,abd\,(\omega/\pi c)^3,$$

not the total volume of the ellipsoid since k, l, and m must all be positive integers. The three amplitudes A_{klm}, B_{klm}, C_{klm} may have any value provided that $A_{klm}k' + B_{klm}l' + C_{klm}m' = 0$. Therefore there are two independent directions of vibration for each set of k, l, m, which doubles the number of possible modes. Hence, the total number of different modes of oscillation up to and including a frequency ω will be $V\omega^3/3\pi^2c^3$, where V

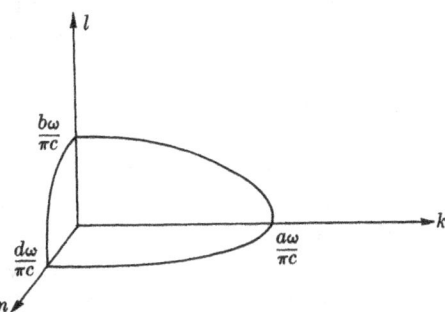

Figure 1.2 *Illustration of the octant of the ellipsoid expressed in Eq.* (1.11) *containing all those values of k, l, m which make the left-hand side of Eq.* (1.11) *less than or equal to the right-hand side. Each mode of oscillation is represented by a single point on the diagram at an integral value of k, l, and m. The dimensions of the graph have been so chosen that each point is equivalent to a unit volume of the ellipsoid.*

is the volume of the cavity. The total possible number of distinct electromagnetic waves or modes of oscillation per unit volume within the cavity, whose frequency is less than or equal to ω, is therefore $\omega^3/3\pi^2c^3$. The number of vibrational modes per unit volume having a frequency between ω and $\omega + d\omega$ is found by differentiating that expression:

$$N(\omega)\,d\omega = \left(\frac{\omega^2}{\pi^2c^3}\right) d\omega \qquad (1.12)$$

or, in cycles per second,

$$N(\nu)\,d\nu = (8\pi\nu^2/3c^3)\,d\nu$$

As long as the total energy in the box is conserved, two of the amplitudes A_{klm}, B_{klm}, C_{klm}, of each single oscillation are in no way restricted. The validity of this important assumption will be questioned later in this chapter.

1.3   THE RADIATION ENERGY DENSITY

The energy in each mode of oscillation is proportional to the square of the amplitude of the particular oscillation. To obtain the energy density inside the box as a function of frequency, Eq. (1.12) must be multiplied by the average or mean energy of each mode of oscillation. To find the average energy of a wave we may note that the electromagnetic vibrations are analogous to mechanical oscillations. By introducing normal coordinates to replace each separate degree of freedom or mode of oscillation one may identify each wave with a corresponding mechanical oscillator having the same frequency. Statistical mechanics is unfortunately outside the scope of this course, but from his classic studies of heat Maxwell was able to show that any particular part of a classical system, such as a single oscillator in a system of oscillators, has a probability of possessing an energy E between E and $E + dE$ proportional to $e^{-E/kT} dE$, where k is the experimentally determined Boltzmann constant, $1.38 \cdot 10^{-16}$ erg/°K, and T is the absolute temperature of the system. The constant of proportionality is determined by requiring that the total probability that this part of the system has any energy at all must be unity. Hence the probability that any, given mode of oscillation will have an energy E between E and $E + dE$ $p(E) \, dE$, will be given by

$$p(E) \, dE = \frac{e^{-E/kT} \, dE}{\displaystyle\int_0^\infty e^{-E/kT} \, dE} = \frac{e^{-E/kT} \, dE}{kT} \tag{1.13}$$

The average or expected energy of each mode of oscillation is obtained by averaging the energy over the probability of having that particular energy:

$$\langle E \rangle = \int_0^\infty E p(E) \, dE = \int_0^\infty \frac{E \, e^{-E/kT} \, dE}{kT} = kT \tag{1.14}$$

Hence the energy density $u(\omega)$ of the modes of oscillation having a frequency between ω and $\omega + d\omega$ is the product of Eqs. (1.12) and (1.14):

$$u(\omega)\, d\omega = \langle E \rangle N(\omega)\, d\omega = kT \left(\frac{\omega^2}{\pi^2 c^3} \right) d\omega \qquad (1.15)$$

Equation (1.15) was first derived by Rayleigh and Jeans and is called the Rayleigh–Jeans law. Unfortunately, as ω increases, the energy density also increases without limit to infinity, leading to an absurd result for infinite frequencies. As shown in Figure 1.3, Eq. (1.15) does indeed fit experimental observations made for very small frequencies or long wavelengths. Observations at high frequencies, however, were empirically described by Wien to conform to

$$u(\omega)\, d\omega = \omega^3\, e^{-\hbar \omega / kT}\, d\omega \qquad (1.16)$$

where $\hbar$ is an experimentally determined constant ($1.05 \cdot 10^{-27}$ erg-sec).

Every step of the derivation of Eq. (1.15), the Rayleigh–Jeans law, resulted from what at the time were well-understood and experimentally confirmed physical principles. Yet it did not fit all of the experimental observations. The trouble lay in assuming that all the modes of oscillation

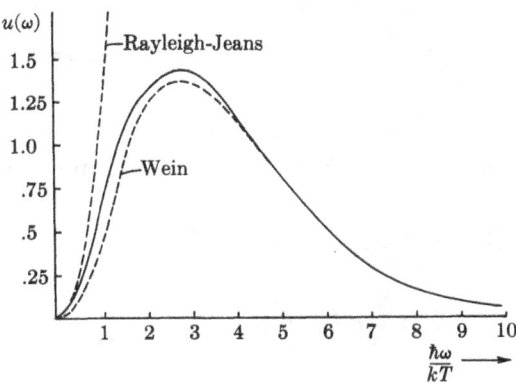

Figure 1.3 *Energy density of the modes of oscillation within a box as a function of frequency. Experimental observations are given by the solid line, the low and high frequency approximations by dotted lines.*

have the average energy kT. Actually, somehow the high frequency modes are frozen out so that their average energy is zero. The solution thus lay in questioning the assumption that the amplitude (and therefore the energy) of each mode of oscillation was unrestricted. If we assume that only discrete values of wave energies are possible, then the average energy of each wave or mechanical oscillator will be different from the constant classical value kT, as we will now show. This is as though one were to assume, for example, that ocean waves enclosed in a breakwater could have heights of the square root of one, two, or three feet but not of one and a half or two and three quarters feet or any other in-between value. If the energy of each wave can be only some integral multiple of $\hbar\omega$, that is,

$$E = n\hbar\omega \qquad (1.17)$$

then Eq. (1.13) must be replaced by a different expression for the probability that any given mode of oscillation will have a unique energy E. Since

$$\sum_{n=0}^{\infty} e^{-n\hbar\omega/kT} = \sum_{n=0}^{\infty} y^n = \frac{1}{1-y} = \frac{1}{1-e^{-\hbar\omega/kT}} \qquad (1.18)$$

where with $x = \hbar\omega/kT$, $y = e^{-x}$, we have for the number of discrete states $p(E)$ having an energy $E = n\hbar\omega$,

$$p(E) = \frac{e^{-n\hbar\omega/kT}}{\displaystyle\sum_{n=0}^{\infty} e^{-n\hbar\omega/kT}} = (1 - e^{-\hbar\omega/kT})\, e^{-n\hbar\omega/kT} \qquad (1.19)$$

The average energy of each mode of oscillation becomes

$$\langle E \rangle = \sum_{n=0}^{\infty} E p(E) = (1 - e^{-\hbar\omega/kT}) \sum_{n=0}^{\infty} n\hbar\omega\, e^{-n\hbar\omega/kT}$$

$$= (1 - e^{-\hbar\omega/kT})\hbar\omega \frac{e^{-\hbar\omega/kT}}{(1 - e^{-\hbar\omega/kT})^2}$$

$$= \frac{\hbar\omega}{e^{\hbar\omega/kT} - 1} \qquad (1.20)$$

This result for the summation can be easily seen if we take the derivative with respect to x of both sides of Eq. (1.18).

For $\hbar\omega \gg kT$, $\langle E \rangle$ is seen to be approximately $\hbar\omega e^{-\hbar\omega/kT}$, which does approach zero as ω increases. This high frequency result for the average energy is readily understood from the first two terms in the summation, $0 + \hbar\omega e^{-\hbar\omega/kT}$, which illustrate that if the first level above $n = 0$ has energy much greater than kT it will not be occupied on the average because $e^{-\hbar\omega/kT} \ll 1$. Multiplying Eq. (1.12), based on Maxwell's theory of electricity and magnetism, by our newly adopted value for the average energy of each mode of oscillation, Eq. (1.20), we find for the energy density of waves having a frequency ω within a range $d\omega$,

$$u(\omega)\,d\omega = \frac{(\hbar\omega^3/\pi^2 c^3)}{e^{\hbar\omega/kT} - 1}\,d\omega \qquad (1.21)$$

which is the Planck formula for the energy distribution for the modes of oscillation within a black body. Throughout this book a bar through a letter means that the physical constant or quantity the letter represents has been divided by 2π; thus $\hbar = h/2\pi$ where h is a natural constant, Planck's constant, experimentally observed to be $6.6252 \pm 0.0005 \times 10^{-27}$ erg-sec. If one expresses the frequency in cycles per second, ν, instead of in radians per second, ω, the energy of a wave would be written $E = nh\nu$ instead of as in Eq. (1.17).

At low frequencies $e^{\hbar\omega/kT}$ is approximately equal to $1 + \hbar\omega/kT$, so that the right-hand side of Eq. (1.20) becomes simply the classical result, kT. At high frequencies the one in the denominator of Eq. (1.21) may be neglected. Thus we should expect the Rayleigh–Jeans law to be valid at low frequencies and Wien's law to be valid at high frequencies, which is in fact the case. Intermediate frequencies can be described only by the exact formula, Eq. (1.21).

It should be emphasized that Planck's use of $E = n\hbar\omega$ (Eq. 1.17) was a theoretical attempt to justify an empirical equation (1.21) which fitted the experimental observations remarkably well; there was no theoretical justification for using (1.17) other than that the experimental observations could be interpreted as requiring (1.17) to be satisfied by the individual electromagnetic waves in the box, or by the fictitious mechanical oscillators' in normal coordinates corresponding to the electromagnetic waves in the box. It is easiest and most straightforward, however, to conclude for the present, subject to further examination later, that the energy transitions of all waves or oscillators occur in quantum jumps. Nondiscrete energy changes for any given frequency or oscillation are somehow forbidden.

In deriving our results for the radiation inside a black body we assumed metallic walls. But it would have made no difference if the walls had been made of any other material. Black body radiation is completely independent of the material walls, and is observed and predicted to be always the same function of temperature provided only that all radiation incident on a pinhole in one wall is either lost through multiple reflections inside the box or, if incident on the pinhole from within the box, is transmitted freely to the exterior.

1.4 SIGNIFICANT MATHEMATICAL PROCESSES USED IN THE DERIVATION

There are several results, incidental to the derivation of Eqs. (1.10), which should be emphasized because of their significance and utility to later developments:

1. If Eq. (1.7) is substituted into Eq. (1.5) there results

$$\nabla^2 \mathbf{E} = -\frac{\omega^2}{c^2} \mathbf{E} \tag{1.22}$$

∇^2 defines an *operation* to be carried out on **E**; the particular operation stated in Eq. (1.22) consists of taking partial derivatives of **E** stated explicitly in Eq. (1.6) in terms of Cartesian coordinates, but in other cases the operation may be multiplication by a constant or by a function of space or time, for example. ∇^2 is an *operator* whose operation on the wave amplitude or wave function **E** results, in this instance, in a constant (namely ω^2/c^2) times the wave function.

2. Since the waves were confined in a box, the boundary conditions (i.e., conditions imposed on the wave function at the walls, which required the tangential component of **E** to be zero in the case of metallic walls) allowed only those solutions for which ω satisfies Eq. (1.11) where k, l, and

m must be integers. This result is a phenomenon common to all confined waves; in vibrating violin strings or organ pipes, for example, it also happens that only those frequencies which satisfy the boundary conditions are permitted.

3. The values of ω^2/c^2 which satisfy the boundary conditions are called characteristic values or more commonly *eigenvalues* of the system. The solutions for E of Eq. (1.22) for particular eigenvalues ω^2/c^2 are called characteristic functions or more commonly *eigenfunctions*. Thus, the operator ∇^2 operating on the particular eigenfunction E_z illustrated in Figure 1.1 has the eigenvalue $17(\pi/a)^2$ in the case of the cube of side a. This particular eigenvalue $17(\pi/a)^2$ has five different nonzero eigenfunctions corresponding to the different k, l, and m consistent with this eigenvalue. An eigenvalue of $16.7(\pi/a)^2$ with its associated eigenfunctions, for example, is not possible for such a system since no combination of integral k, l, and m can give such an eigenvalue. (If we include the E_x and the E_y solutions, as we must, a total of eleven different eigenfunctions are possible for $\omega^2/c^2 = 17(\pi/a)^2$.)

Problem 1.1 Find the four lowest eigenvalues and their corresponding eigenfunctions of a vibrating violin string 10 cm long if the tension on the string, T, is 10 newtons and the string has a mass per unit length, μ, of 0.2 kg/m. The wave equation for transverse waves in a string is

$$\frac{\partial^2 S}{\partial x^2} = \frac{\mu}{T}\frac{\partial^2 S}{\partial t^2}$$

where S is the perpendicular displacement of the string from equilibrium, x is the distance along the string, and t is the time. (Include the possibility that S may be taken in each of two perpendicular directions.)

Problem 1.2 Graph the Rayleigh–Jeans law and the Planck black body distribution as a function of frequency for the surface of the sun, whose temperature is 6000°K. Be careful in labeling the ordinate to give the units and magnitude of what is observed from a black body surface. On the same graph paper plot the average energy of the electromagnetic waves on the solar surface as a function of their frequency. (Surface emission per unit area is proportional to the velocity of light, c, times the energy density per unit volume, $u(\omega)$.)

Problem 1.3   What are the minimum energies electromagnetic waves of wavelengths 5000 Å and 10 m may have?

Problem 1.4   What are the eigenvalues and eigenfunctions of the operator $(i\ d/dx)^2$ if the eigenfunctions are required to be zero when $x = 0$ and 2?

BIBLIOGRAPHY

Suggested books for further reading:

Bergmann, P. G., *Basic Theories of Physics*, Vol. II. Englewood Cliffs, N.J.: Prentice-Hall Inc., 1951. This book contains an excellent and detailed treatment of black body radiation and the early quantum theory on which much of this chapter is based. The author is particularly gifted in presenting apt examples and models in his lucid presentation.

Slater, J. C., *Modern Physics*. New York: McGraw-Hill Book Co., 1955. This book is notable for its exceptionally clear and interesting exposition.

IN DESCRIBING THE ELECTRIC FIELD VECTOR everywhere within the box (the black body cavity) at a given instant of time through Eq. (1.10), we obtained an alternative way of specifying the condition of the radiation field within the box. It is equally possible to specify the electric field vector as an explicit function of the spatial coordinates x, y, and z, or alternatively to specify the amplitude of each standing wave, as is suggested by Eq. (1.10). Either description serves to completely *represent* the physical *state* of the interior of the box. The method of representing a state of radiation in terms of its various modes of oscillation is developed more generally in this chapter.

The customary representation in terms of spatial coordinates is replaced by alternative representations in terms of modes, which are of great utility in computing and understanding quantum mechanical behavior. It will be shown, by representing a wave in terms of its frequency dependence as well as in terms of its time dependence, that *one cannot simultaneously measure the frequency of a wave and the time at which the wave is absorbed in a detector with unrestricted precision.* This imprecision or *uncertainty* is a necessary consequence of all wave motion and is of central importance to the quantum theory. Indeed, Heisenberg originally derived quantum mechanics by postulating this uncertainty as fundamental to all physical measurement. In later chapters it will transpire that a description of any physical system can be made only in terms of pairs of *complementary* physical variables, both of which cannot be measured simultaneously with infinite precision. In this chapter it will be shown that it is impossible to limit the range of frequency or wave number *and* the temporal duration or spatial position of a wave; in the following chapter, in addition to discussing the complementary variables of time and frequency or energy, we will take up the physical impossibility of simultaneously measuring position and momentum with infinite precision.

Sines and cosines are good examples of functions whose product with any other function of the same class gives zero when integrated over all ranges of the variable, unless the two multiplied functions are identical. Such sets of functions are said to be orthogonal. For example, the integrals over a period of the fundamental frequency ω_0 of the products $\sin m\omega_0 t \sin n\omega_0 t$, $\sin m\omega_0 t \cos n\omega_0 t$, and $\cos m\omega_0 t \cos n\omega_0 t$, where m and n are integers, are

$$\frac{\omega_0}{2\pi} \int_0^{2\pi/\omega_0} \sin m\omega_0 t \sin n\omega_0 t \, dt$$

$$= \frac{\omega_0}{2\pi} \left[\int_0^{2\pi/\omega_0} \frac{\sin (m-n)\omega_0 t}{2(m-n)\omega_0} - \frac{\sin (m+n)\omega_0 t}{2(m+n)\omega_0} \right] = 0 \qquad \text{if } m \neq n$$

$$= \frac{\omega_0}{2\pi} \int_0^{2\pi/\omega} \sin^2 m\omega_0 t \, dt = \frac{\omega_0}{2\pi} \int_0^{2\pi/\omega_0} (\tfrac{1}{2} - \tfrac{1}{2} \cos 2m\omega_0 t) \, dt = \tfrac{1}{2} \qquad \text{if } m = n$$

$$\frac{\omega_0}{2\pi} \int_0^{2\pi/\omega_0} \sin m\omega_0 t \cos n\omega_0 t \, dt$$

$$= -\frac{\omega_0}{2\pi} \left[\int_0^{2\pi/\omega_0} \frac{\cos (m-n)\omega_0 t}{2(m-n)\omega_0} + \frac{\cos (m+n)\omega_0 t}{2(m+n)\omega_0} \right] = 0 \qquad \text{if } m^2 \neq n^2$$

$$= \pm \frac{1}{2} \frac{1}{m\omega_0} \frac{\omega_0}{2\pi} \left[\int_0^{2\pi/\omega_0} \sin^2 m\omega_0 t \right] = 0 \qquad \text{if } m^2 = n^2$$

and similarly

$$\frac{\omega_0}{2\pi} \int_0^{2\pi/\omega} \cos m\omega_0 t \cos n\omega_0 t \, dt = \tfrac{1}{2}\delta_{mn}$$

where δ_{mn}, called a Kronecker delta, is zero for $m \neq n$ and unity for $m = n$. Besides sines and cosines, other examples of orthogonal functions that we shall encounter are Bessel functions and Legendre polynomials. The coefficients of such functions may frequently be used to describe a given function of coordinates or some other variable, so that such functions furnish a new and frequently more convenient kind of "space" in which to describe the given function.

2.1 FOURIER SERIES

Suppose that a given oscillation or periodic function of time, $f(t)$, with period $2\pi/\omega_0$, is to be described by a sum or superposition of many waves, specifically an infinite sine and cosine series, called a Fourier series:

$$f(t) = \sum_{n=0}^{\infty} A_n \sin n\omega_0 t + B_n \cos n\omega_0 t \qquad (2.1)$$

where the n's are integers, ω_0 is a given fundamental frequency, and the A's and B's are the amplitudes of each component frequency. Multiplying both sides of Eq. (2.1) by $\sin m\omega_0 t$ and integrating over the fundamental period, we find that

$$\int_0^{2\pi/\omega_0} f(t) \sin m\omega_0 t \, dt \qquad (2.2)$$

$$= \sum_{n=0}^{\infty} \int_0^{2\pi/\omega_0} A_n \sin n\omega_0 t \sin m\omega_0 t \, dt + B_n \int_0^{2\pi/\omega_0} \cos n\omega_0 t \sin m\omega_0 t \, dt$$

The sines and cosines are orthogonal functions, and all the terms of the infinite series in n on the right-hand side of Eq. (2.2) give zero except for the one term in the sine series for which $n = m$, which gives

$$\int_0^{2\pi/\omega_0} f(t) \sin m\omega_0 t \, dt = \frac{\pi}{\omega_0} A_m \qquad (2.3)$$

Similarly by multiplying (2.1) by cos $m\omega_0 t$,

$$B_m = \frac{\omega_0}{\pi} \int_0^{2\pi/\omega_0} f(t) \cos m\omega_0 t \; dt \qquad (2.4)$$

As in the preceding, our function may be given as an explicit function of time, $f(t)$; alternatively, however, we may compute the A's and B's from a knowledge of $f(t)$ and use them to describe our function:

$$f(t) = \sum_{n=0}^{\infty} \left\{ \frac{\omega_0}{\pi} \left[\int_0^{2\pi/\omega_0} f(t) \sin n\omega_0 t \; dt \right] \sin n\omega_0 t \right.$$

$$\left. + \frac{\omega_0}{\pi} \left[\int_0^{2\pi/\omega_0} f(t) \cos nw_0 t \; dt \right] \cos n\omega_0 t \right\}$$

where the quantities in brackets are functions only of frequency, $n\omega_0$, since the time will have been integrated out. In this way $f(t)$ may described, for example, as being made up of oscillations having frequencies $n\omega_0$, each of whose relative amplitudes A_n and B_n represent the quantity just as completely as $f(t)$ does. There is one essential difference, however, in that while all values of t are possible in the time representation, only certain values of frequency, integral multiples of ω_0, are possible in the frequency representation. If $f(t)$ is not periodic this difference disappears, as will be shown in the following section.

As an example, let us consider a sixty-cycle alternating current of unit amplitude, $f(t) = \sin 2\pi 60 t$. If we attempt to find a series to describe it using the basic frequency $\omega_0 = 2\pi 60$, we immediately find the trivial result $A_1 = 1$. All other coefficients are zero. For graphs of alternative representations of 60-cycle a.c. current see Figure 2.1. Note that in this example, the Fourier series consists of only one term because we implicitly assumed that the wave started at $t = -\infty$ and continued to $t = +\infty$. Actually, every time an electric switch is turned on or off higher frequencies are introduced, called transients (see Problem 2.2). These are frequently picked up by household radios as sharp audible clicks.

For a second example, suppose that at

$$0 < t < t_0, \qquad f(t) = 0$$
$$t_0 < t < 2t_0, \qquad f(t) = 1$$

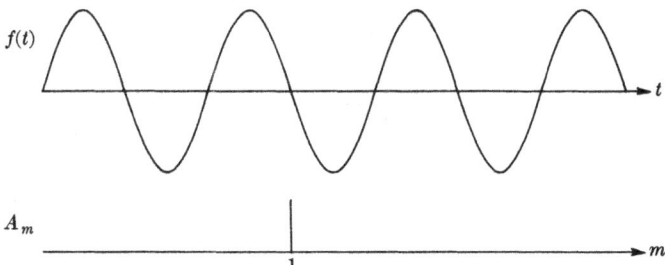

Figure 2.1 *Alternative representations of 60-cycle a.c.; the second graph illustrates the fact that only one frequency is represented.*

and then the function is repeated for larger and smaller integral multiples of t_0. Obviously, a convenient fundamental frequency will be π/t_0:

$$
\begin{aligned}
A_n &= \frac{1}{t_0} \int_0^{t_0} (0) \sin n \frac{\pi}{t_0} t \, dt + \frac{1}{t_0} \int_{t_0}^{2t_0} (1) \sin n \frac{\pi}{t_0} t \, dt \\[2mm]
&= 0 - \frac{1}{t_0} \frac{t_0}{n\pi} \left[\cos n \frac{\pi}{t_0} 2t_0 - \cos n \frac{\pi}{t_0} t_0 \right] \\[2mm]
&= \frac{2}{\pi n} \quad \text{for odd } n; \quad = 0 \quad \text{for even } n.
\end{aligned}
\tag{2.6}
$$

Similarly,

$$
B_n = 0 \quad \text{for} \quad n > 0, \quad B_0 = 1,
$$

and hence

$$
f(t) = \sum_{m=0}^{\infty} \left[1 + \frac{2}{\pi(2m+1)} \sin (2m+1) \frac{\pi}{t_0} t \right]
\tag{2.7}
$$

Alternate representations of this system are presented in Figure 2.2.

Problem 2.1 What are the amplitudes of the various wavelengths present in a vibrating violin string of length l plucked initially to form an isoceles triangle?

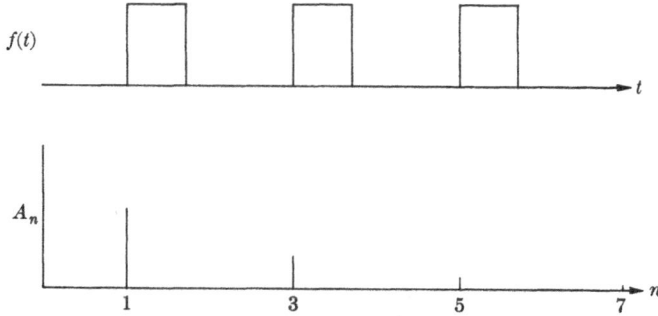

Figure 2.2   *Alternative representations of a periodic step function pulse.*

Problem 2.2   What are the amplitudes of the various frequencies present in half-wave rectified 60-cycle a.c., that is, $0 < t < 1/120$, $f(t) =$ $\sin 2\pi 60t$, and $1/120 < t < 1/60$, $f(t) = 0$, and then the function repeats?

Problem 2.3   What are the amplitudes of the various frequencies present when 60-cycle a.c. is repeatedly turned on for $1/10$ second and then turned off for $4/10$ of a second?

Problem 2.4   Find the two Fourier series for $f(x) = x$ and $|x|$ for $-l < x < l$ ($f(x)$ repeats for other values of x); then compare the series for $0 < x < l$.

Problem 2.5   Find the Fourier series for $f(x) = \cos \mu x$ for $0 < \mu < 1$, $-\pi < x < \pi$ ($f(x)$ repeats for other values of x).

Problem 2.6   Find the two Fourier series for $f(x) = x^2$ and x^3, over $-l < x < l$ ($f(x)$ repeats for other values of x).

2.2   FOURIER TRANSFORMS

The preceding discussion was devoted to alternative representations of functions varying periodically with time. The periodicity resulted in a discrete frequency representation, that is, *Fourier series* composed of terms having frequency some *integral* multiple of the fundamental frequency.

In order to represent the larger class of functions which are not necessarily periodic, we now discuss the case in which the fundamental period approaches infinity and instead of a discrete set of frequencies we have a continuum of oscillation frequencies. We need a new and very important concept, that of the Dirac delta function which is zero everywhere except for one point where it is infinite. Consider the integral of the function $(\sin bu)/u$. It is a standard definite integral whose value is π when integrated over u from $-\infty$ to ∞. For large values of b, however, the sine function oscillates very rapidly as a function of u as shown in Figure 2.3; hence the integral

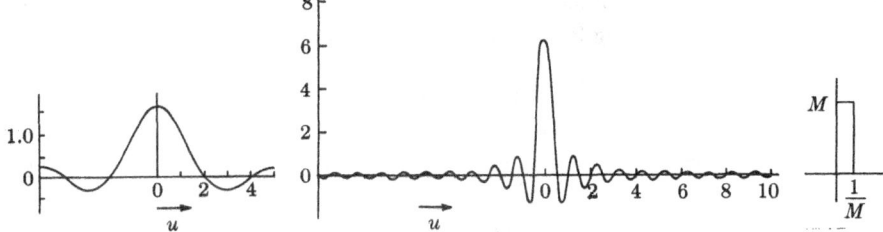

Figure 2.3   *Illustration of the function* $(\sin bu)/u$. *The first figure is for* $b = \pi/2$, *the second for* $b = 2\pi$; *the third function approaches the Dirac delta function as* M *approaches infinity.*

is close to zero between limits of large values of bu of the same sign. As $b \to \infty$ the integral approaches zero, unless the origin is included in the integration interval, in which case the value of the integral remains π, and as the maximum of this function increases, its width decreases in such a way that the total area under the curve is a constant. That is, the integral of this function has a nonzero value only at the origin; such a function is called a Dirac delta function, and is represented by the symbol $\delta(t)$. By definition, $\delta(t) = 0$ except at the point $t = 0$ where it is infinite in such a way that its integral is unity. It may be thought of as a function of infinitesimal width and infinite height such that the area under its curve is one, as is illustrated in the third diagram in Figure 2.3. If the integrand is multiplied by any more slowly varying function of u, say $f(t + u)$, the value of this function is only important at the origin where the integral has a nonzero value. That is:

$$\int_{-a}^{a} f(t+u) \frac{\sin bu}{u} \, du \underset{b\to\infty}{\overset{\lim}{=}} \pi f(t) \tag{2.8}$$

and since

$$\frac{\sin bu}{u} = \int_{0}^{b} \cos u\omega \, d\omega \tag{2.9}$$

$$f(t) \underset{b\to\infty}{\overset{\lim}{=}} \frac{1}{\pi} \int_{-a}^{a} f(t+u) \, du \int_{0}^{b} \cos u\omega \, d\omega \tag{2.10}$$

where $\underset{b\to\infty}{\lim}$ is read "equals in the limit as $b \to \infty$." Interchanging the order of integration and letting first a then b go to infinity, we obtain

$$f(t) = \frac{1}{\pi} \int_{0}^{\infty} d\omega \int_{-\infty}^{\infty} f(t+u) \cos u\omega \, du$$

Let

$$u = v - t \tag{2.11}$$

then

$$f(t) = \frac{1}{\pi} \int_{0}^{\infty} d\omega \int_{-\infty}^{\infty} f(v) \cos \omega(v-t) \, dv$$

Since the cosine is an even function, we may extend the integral over ω to $-\infty$ and multiply by $1/2$. Since the sine is an odd function,

$$f(v)i \sin \omega(v-t)$$

may then be added to the above integral over v without changing the value of the integral, since the integral over ω of this added term is zero. Therefore,

$$f(t) = \frac{1}{2\pi} \int_{-\infty}^{\infty} d\omega \int_{-\infty}^{\infty} f(v) \, e^{-i\omega(t-v)} \, dv \tag{2.12}$$

Thus

$$f(t) = \frac{1}{\sqrt{2\pi}} \int_{-\infty}^{\infty} F(\omega) \, e^{-i\omega} \, d\omega \tag{2.13}$$

where

$$F(\omega) = \frac{1}{\sqrt{2\pi}} \int_{-\infty}^{\infty} f(v) \, e^{i\omega v} \, dv \tag{2.14}$$

$F(\omega)$ is called the Fourier transform of $f(t)$, and is obtained from (ft) by Eq. (2.14), where v is a dummy variable of integration. Conversely

$f(t)$ may be obtained from $F(\omega)$ by the inverse Fourier transform, Eq. (2.13). Hence, if t were to represent time and ω frequency, a particular oscillation or wave could be equally well represented as a function of time, $f(t)$, or as a function of frequency, $F(\omega)$. Similarly in later chapters we will frequently find it convenient to transform from a spatial coordinate representation in ordinary space to a wave number or momentum representation in which momentum forms an equally valid space or set of continuous variables in terms of which the system may be explicitly described.

Equations (2.13) and (2.14) define the Fourier exponential transforms. The analogous Fourier cosine and sine transforms may be obtained from a similar derivation. Alternative spherical harmonic or Bessel integral transform representations may be obtained for spherical or cylindrical waves, for example, but our discussion in this chapter will be limited to Fourier exponential transforms.

2.3 RELATION BETWEEN THE WIDTH OF A FUNCTION AND ITS TRANSFORM

For our trivial sixty-cycle a.c. current of unit amplitude for example, we have $F(\omega) = \sqrt{2\pi}\delta(\omega - 2\pi 60)$, for only $\omega = 2\pi 60$ is present, as may be readily confirmed by evaluating Eq. (2.14) for $f(t) = e^{-i2\pi 60t}$. Conversely, from Eq. (2.13), $f(t) = e^{-i2\pi 60t}$ for $F(\omega) = \sqrt{2\pi}\delta(\omega - 2\pi 60)$. If the wave were not of infinite duration, however, for instance if a light switch were rapidly flicked on and off, $F(\omega)$ would no longer be a delta function consisting of only one frequency. Suppose the switch flicked on at time $-t_0$ and off at time $+t_0$; then from Eq. (2.14) we find

$$
\begin{aligned}
F(\omega) &= \frac{1}{\sqrt{2\pi}} \int_{-t_0}^{t_0} e^{-i2\pi 60t}\, e^{i\omega t}\, dt \\[2mm]
&= \frac{1}{\sqrt{2\pi}} \frac{e^{i(\omega - 2\pi 60)t_0} - e^{-i(\omega - 2\pi 60)t_0}}{i(\omega - 2\pi 60)} \\[2mm]
&= \frac{2}{\sqrt{2\pi}} \frac{\sin(\omega - 2\pi 60)t_0}{\omega - 2\pi 60}
\end{aligned} \tag{2.15}
$$

which we recognize from our earlier discussion to be a delta function times $\sqrt{2\pi}$ as t_0 increases without limit. For finite t_0, however, $F(\omega)$ is not a delta function but contains frequencies other than $2\pi 60$, as is shown in Figure 2.4.

From this example one can see that the range of frequencies present in a finite wave train is, at least in this instance, inversely related to the

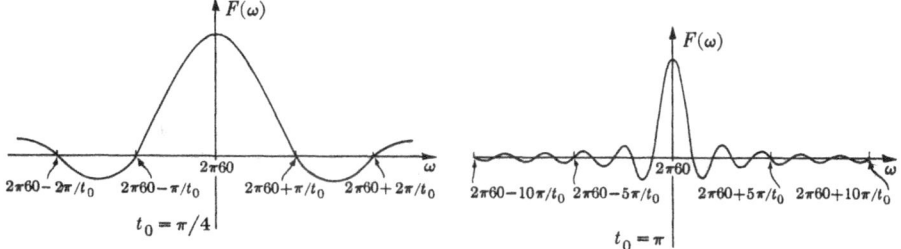

Figure 2.4 *The Fourier transform of a wave train of frequency $2\pi 60$ which is turned on for a time $2t_0$. Note that the frequency $2\pi 60 + (5\pi/25t_0)$, for example, is present with amplitude as much as 13% of the amplitudes of $2\pi 60$ itself. The figure on the left is for $t_0 = \pi/4$; the figure on the right is for larger $t_0 = \pi$. The ordinate and abscissa scales of the two figures are the same.*

duration of the wave train. This is not at all a mathematical trick; an apparatus such as a spectroscope employed to detect any of these frequencies in such a finite wave train would yield a positive result.

The Fourier transform of a Gaussian distribution function in t is a Gaussian distribution function in ω. Specifically, if

$$f(t) = \frac{A}{\sqrt{2}} e^{-t^2/2(\Delta t)^2} \tag{2.16}$$

is inserted into Eq. (2.14) there results

$$F(\omega) = \frac{A}{\sqrt{4\pi}} \int_{-\infty}^{\infty} e^{-t^2/2(\Delta t)^2 + i\omega t}\, dt$$

$$= A\Delta t\, e^{-\omega^2(\Delta t)^2/2} \tag{2.17}$$

Thus, a Gaussian distribution in time with a root mean square deviation in time of Δt leads to a Gaussian distribution in frequency with a root mean square deviation in frequency of $\Delta\omega = 1/\Delta t$. The inverse relationship between spread in time of a wave and the spread in frequency is a very general one, as may be inferred from Eqs. (2.13) or (2.14). If $f(v)$ is not oscillatory and peaks at $v = 0$, the integrand in Eq. (2.14) represents a sum of waves $e^{i\omega v}$ of differing v with weight $f(v)$. The waves will interfere constructively only if their phase differs by less than π. Hence, a broad maximum in v would result in destructive cancellation except near $\omega = 0$, whereas a sharply peaked distribution in v would result in little destructive interference over a large spread in $\omega \neq 0$.

We must conclude that the product of the spread in frequency of a pulse of radiation, $\Delta\omega$, and the duration of the pulse, Δt, is quite generally of the order of unity. That is,

$$\Delta\omega\Delta t \sim 1 \qquad (2.18)$$

or remembering the relation between the least unit of radiation energy and frequency stated in the previous chapter (Eq. 1.17), $E = \hbar\omega$,

$$\Delta E\Delta t \sim \hbar \qquad (2.19)$$

The relationship between frequency and time expressed in Eq. (2.18) is a fundamental characteristic of all wave motion.

A wave train whose duration is limited is necessarily comprised of a range of frequencies. A single cycle of high frequency music has a very short duration and requires a large range of AM broadcast frequency to carry it. Thus some good music stations are assigned frequencies at the extreme high end of the AM radio broadcast band (so high that not all older receiving sets were built to detect them and the broadcaster loses a potential audience) both because at these frequencies a wider bandwidth, 20 kc, can be received in most AM radios (since the bandwidth divided by the frequency is nearly constant in most AM sets) and because they require an unpopular broadcast frequency since for higher fidelity they have requested a bandwidth of 20 kc, twice the frequency spread normally assigned. The higher the audio frequency with which the amplitude of the broadcast frequency is modulated, the more rapid must be the changes in this amplitude. This rapid modification requires the availability of a finite range of frequencies above and below the assigned broadcast frequency. Hence, Eq. (2.18) requires broader frequency assignments for higher fidelity in AM broadcasting, which may be represented approximately by $\Delta\nu/\Omega \gtrsim 1$ where $\Delta\nu$ is the assigned spread in broadcast frequency (ordinarily 10 kc) and Ω is the highest reproducible audio frequency for the assigned AM band width. Of course, also the AM receiver must not be too selective of broadcast frequency, in order to make full use of the high fidelity transmission.

Problem 2.7 What is the frequency spectrum of the intensity of a flash of lightning which might be represented as a sudden magnetic field decreasing exponentially with time as $e^{-t/t}$? That is, compute $F(\omega)$ and then the intensity, $I = F^*(\omega)F(\omega)$, where $F^*(\omega)$ is the complex conjugate of $F(\omega)$. What is the frequency spread in which the intensity falls to one half its peak value if $t_0 = 10^{-2}$ sec? if $t_0 = 1$ sec? (These frequencies show up as whistles in unselective AM radio receivers, due to the difference in propagation velocity of the various frequencies in the ionosphere. Thus a lightning stroke in South America produces an electromagnetic disturbance which travels along a magnetic field line out into the ionosphere and back into the United States. The higher frequencies arrive first, the low last, and one hears a high note which drops rapidly in pitch, that is, a whistle in the receiver.)

2.4 WAVE PACKETS

A plane wave propagated along the x axis may be represented by the expression

$$E = E_0\, e^{i(k_0 x - \omega t)} \tag{2.20}$$

where k_0 is the wave number, defined in terms of the wavelength,

$$k_0 = 1/(\lambda_0/2\pi) = 1/\bar\lambda_0$$

The phase, $k_0 x - \omega t$, remains constant for $x = (\omega/k_0)t = \omega\bar\lambda_0 t$. Hence, the phase of the wave propagates with a velocity $\omega\bar\lambda_0$. Therefore, if the wave represents light propagating in a medium, ω may be written as $c/n\bar\lambda_0$ where c is the phase velocity of the wave *in vacuo* (2.998 · 10^8 m/sec for electromagnetic radiation) and n is the ratio of the phase velocity *in vacuo* to the phase velocity in the propagating medium. In general n is some function of the wavelength of the radiation and ω is not a linear function of k.

The previous discussion concerning the complementary variables ω and t may be taken over entire for the one-dimensional variables k and x, which are also complementary. The formal mathematical treatment is identical; by taking the Fourier transform of $f(x)$ one may go to k space, just as in the previous discussion one transformed from the variable t to ω:

$$F(k) = \frac{1}{\sqrt{2\pi}} \int_{-\infty}^{\infty} f(x)\, e^{ikx}\, dx \qquad (2.21)$$

If one wishes to represent a pulse of light, that is, a bundle of radiation or a *wave packet* which propagates through the medium, a monochromatic wave such as is represented by Eq. (2.20) will not do. Instead, one requires

$$E(x,t) = f(x,t) e^{ik_0 x} \qquad (2.22)$$

Assuming the wave packet centered at $x = 0$ when $t = 0$, $f(x,0)$ is a function which is large near $x = 0$ and small at x far from the origin. Alternatively $E(x,t)$ can be expressed in terms of $F(k',t)$, the x-space Fourier transform of $f(x,t)$,

$$F(k',t) = \frac{1}{\sqrt{2\pi}} \int_{-\infty}^{\infty} f(x,t)\, e^{ik_0 x}\, e^{ik'x}\, dx = \frac{1}{\sqrt{2\pi}} \int_{-\infty}^{\infty} f(x,t)\, e^{ikx}\, dx \equiv F(k - k_0, t)$$

where $k = k' + k_0$. Thus $F(k - k_0, t)$ is a distribution in wave number, rather than a delta function in k representing a monochromatic wave (whose inverse transform would give Eq. (2.20) instead of the desired Eq. (2.22)).

We illustrate the behavior of a wave packet, using a Gaussian distribution function again, because of the convenience of its simple transform properties. Suppose that

$$F(k - k_0, t) = B\, e^{-(k-k_0)^2/2(\Delta k)^2}\, e^{-i\omega t} \qquad (2.23)$$

Equation (2.23) is the weighting function of a group of waves whose integral over k (the Fourier transform) will yield $E(x,t)$:

$$E(x,t) = \frac{1}{\sqrt{2\pi}} \int_{-\infty}^{\infty} B\, e^{-(k-k_0)^2/2(\Delta k)^2}\, e^{i(kx-\omega t)}\, dk \qquad (2.24)$$

Since ω is a function of k one must include it in the integral, and hence one can solve the integral exactly only for specific examples of $\omega(k)$. Fortunately, if $F(k - k_0)$ is a reasonably narrow distribution function, one may solve (2.24) in terms of the derivatives of ω to good approximation. In this case ω is expanded in a Taylor's series about $k = k_0$,

$$\omega(k) = \omega(k_0) + \left(\frac{\partial \omega}{\partial k}\right)_{k=k_0} (k - k_0) + \frac{1}{2}\left(\frac{\partial^2 \omega}{\partial k^2}\right)_{k=k_0} (k - k_0)^2 \quad (2.25)$$

The first derivative with respect to k has a particularly useful and simple physical interpretation. In a dispersive medium, n and therefore the phase velocity of a wave depend on the wavelength, and one cannot define a single phase velocity for the entire group of waves or wave packet given by Eq. (2.24). In detecting a wave packet one observes the intensity of the wave. One receives a maximum signal when the maximum amplitude of the wave arrives. Under these circumstances one does better to find the velocity of the maximum of the wave packet and call it the *group velocity*. The maximum of the wave packet in momentum space will occur for $k = k_0$ (because $F(k - k_0)$ is a maximum at $k = k_0$). The wave packet will have a maximum in coordinate space as well, at that position where the least rapid phase variation with k occurs, yielding maximal constructive interference in the integral given in Eq. (2.24) for the wave amplitude. That is, the extremum in the phase of the integrand in Eq. (2.24) must be sought:

$$\frac{\partial}{\partial k}(kx - \omega t) = x - \frac{\partial \omega}{\partial k} t = 0 \quad (2.26)$$

The *position* of maximal constructive interference is seen therefore to propagate with a velocity $\partial \omega / \partial k$ called the *group velocity* v_g.

The difference between phase and group velocity of a wave packet may be illustrated by a rotating barber pole with helical stripes moving with a translational velocity parallel to its axis. Even when the "group velocity" (translational velocity) of the pole is zero, a colored portion of the pole is observed to "move" if the pole rotates, with a "phase velocity" given by $s\nu$ where s is the pitch of the stripes (distance along the pole that a stripe moves in one rotation) and ν is the number of rotations per second. The phase velocity would be different if the pole moved with a velocity v_g, namely, it would be $v_g + s\nu$.

Another frequently cited example of differing phase and group velocities is that of wavelets caused by a stone dropped in water. The outermost ring of ripples propagates outward with a group velocity v_g, but ripples within this outer ring are observed to propagate with a higher velocity and to disappear when they reach the outer, slower moving edge of the ripples.

Labeling $\partial\omega/\partial k$ as v_g and $\partial^2\omega/\partial k^2$ as a, substituting Eq. (2.25) into Eq. (2.24), and letting $k' = k - k_0$, we obtain

$$
E(x,t) = \frac{B}{\sqrt{2\pi}} \exp\{i[k_0x - \omega(k_0)t]\} \exp\left\{-\frac{1}{2}\frac{(x - v_gt)^2}{(\Delta k)^{-2} + iat}\right\}
$$

$$
\times \int_{-\infty}^{\infty} \exp\left\{-\frac{1}{2}[(\Delta k)^{-2} + iat]\left[k' - \frac{i(x - v_gt)^2}{(\Delta k)^{-2} + iat}\right]^2\right\} dk'
$$

$$
= \frac{B}{\sqrt{(\Delta k)^{-2} + iat}} \exp\{i[k_0x - \omega(k_0)t]\} \exp\left\{-\frac{1}{2}\frac{(x - v_gt)^2}{(\Delta k)^{-2} + iat}\right\}
$$

$$
= B\left[\frac{(\Delta k)^{-2} - iat}{(\Delta k)^{-4} + a^2t^2}\right]^{1/2} \exp\left\{i\left[k_0x - \omega(k_0)t + \frac{at(x - v_gt)^2}{2[(\Delta k)^{-4} + a^2t^2]}\right]\right\}
$$

$$
\times \exp\left\{-\frac{1}{2}\frac{(x - v_gt)^2(\Delta k)^{-2}}{(\Delta k)^{-4} + a^2t^2}\right\} \tag{2.27}
$$

which is illustrated in Figure 2.5.

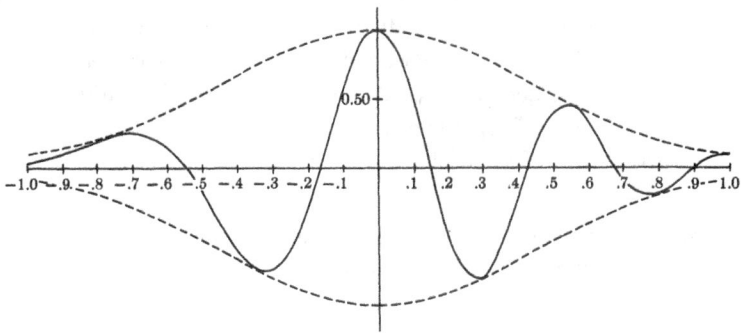

Figure 2.5 *The real part of the amplitude of a wave packet as a function of spatial coordinate* x, *which has a Gaussian distribution in* k *and* x. *The dashed line is the Gaussian envelope of the packet.*

The observed intensity of the wave packet is the square of the absolute value of the wave amplitude $E(x,t)$,

$$I = E(x,t)E^*(x,t) = \frac{B^2}{\sqrt{(\Delta k)^{-4} + a^2 t^2}} \exp\left\{ -\frac{(x - v_g t)^2 (\Delta k)^{-2}}{(\Delta k)^{-4} + a^2 t^2} \right\} \quad (2.28)$$

The expression on the right-hand side of Eq. (2.28) is simply a Gaussian function in x about a mean value of $v_g t$ whose root mean square deviation in x, Δx, is given by

$$\Delta x = \frac{\sqrt{(\Delta k)^{-4} + a^2 t^2}}{(\Delta k)^{-2}} = \frac{1}{\Delta k} \sqrt{1 + a^2 t^2 (\Delta k)^4} \quad (2.29)$$

Hence a wave packet in a dispersive medium ($a \neq 0$) spreads out in time, as would be expected since then different elements of the wave packet (waves of differing frequency) propagate with different velocities.

Problem 2.8   Radio waves in the ionosphere propagate with an index of refraction given approximately by

$$n = 1 + \frac{\omega_p^2/2}{\omega_p^2 - \omega^2 + [\gamma^2/(\omega_p^2 - \omega^2)]}$$

where ω_p is the natural frequency of electron oscillations, equal to a maximum of 5×10^7 radians/sec in the F layer where the electron density is a maximum. If $\gamma = 10^{15}$ (radians/sec)^2, what are the phase and group velocities of a wave of frequency 3×10^7 radians/sec? 7×10^7 radians/sec? Within how great a distance will a pulse 10^4 meters long, with $\omega_0 = 3 \times 10^7$ radians/sec, have lengthened 10%? (A pulse of this length would be required to broadcast high audio frequencies and a change in the length would result in distortion.) (Hint: First find the *time* in which the pulse lengthens by 10%.)

Problem 2.9   The velocity of crests of gravitational waves on water is related to their wavelength λ as follows:

$$v = \sqrt{g\lambda}$$

where g is 980 cm/sec^2. What is the group velocity?

Problem 2.10   If the group velocity of a wave is given by $v_g = 3v_p$ where v_p is the phase velocity, how does v_p depend on ω?

BIBLIOGRAPHY

Bohm, David, *Quantum Theory.* Englewood Cliffs, N.J.: Prentice-Hall Inc., 1951. This book contains an exceptionally fine and detailed discussion of wave packets and their transform properties. See Chapter 3.

Jenkins, F. and H. White, *Fundamentals of Optics.* New York: McGraw-Hill Book Co., 1950.

Rossi, Bruno, *Optics.* Reading, Mass.: Addison-Wesley, Inc., 1957. See Chapter 5 for a discussion of the velocity of propagation of waves.

DIFFRACTION AND INTERFERENCE

OF PHOTONS AND PARTICLES

3

FROM OUR STUDY of black body radiation we saw that Planck obtained a satisfactory theoretical prediction of black body radiation by assuming, instead of an arbitrary energy of oscillation, that the modes of oscillation in a black body could possess only discrete amounts of energy. The basis for this assumption was no more nor less than that the experimentally observed spectrum may be interpreted in this way. This discrete, quantized behavior of electromagnetic radiation strongly suggests that light waves are transmitted as packets or quanta of energy. This chapter will be devoted to exploring further particle-like characteristics of electromagnetic waves and the wave-like characteristics of particles.

Atoms emit and absorb light energy in discrete amounts or quanta, as was initially pointed out by Einstein in his interpretation of the photoelectric effect and has subsequently been confirmed by many diverse experimental observations. The light is emitted and absorbed, in the volume of a single atom, as light quanta called photons. The position and motion of the photons are determined by means of wave packets localized in space and time but much less precisely localized than the atoms themselves. Our discussion of diffraction and interference phenomena in terms of wave packets will prove very suggestive and illuminating for an understanding of diffraction and interference phenomena of particles, which phenomena in turn require an inherent space-time localization as the essence of the "common sense" model of a particle.

This chapter will further review those aspects of particle behavior which cannot be explained by a model of particle behavior requiring infinitely precise space-time localization. The uncertainty or indeterminacy in complementary variables, such as frequency (or energy) and time, which is a fundamental characteristic of wave motion, will be shown to characterize particle behavior as well. The discreteness or noncontinuity of available values of normally continuous physical observables of radiation (for example, the frequency of a confined wave) is observed also for closely confined particles (for example, the energy of electrons in an atom). Finally, diffraction and interference phenomena, which are the sole basis for the credibility of the wave theory of light, will be discussed for particle experiments where they are also realized.

3.1 THE PHOTOELECTRIC EFFECT

When light is incident on the atoms of a gas, a liquid, or a solid, electrons are instantaneously emitted. The remarkable feature about this phenomenon, called the photoelectric effect, is that the energy of the electrons is observed to be unaffected by changes in the *intensity* of the incident radiation. An increase in light intensity only increases the number of electrons released. However, the kinetic energy of the electrons is observed to decrease when the *frequency* of the incident radiation is lowered. Indeed, no photoelectrons whatever are observed if the frequency is less than some critical value unique for each substance.

It is impossible to explain these observations by means of the wave theory of light. The wave theory of light leads one to anticipate that a long radio wave incident on an atom could cause enough energy to be absorbed (over sufficiently long radiation times) for an electron to be released.

Moreover, when electrons are emitted, an increase in radiated power for a particular wave should cause an emitted electron to have more kinetic energy rather than more electrons of the same average energy to be emitted. One would further expect the rate of electron emission to increase during the radiation time, being greatest at the end of the radiation pulse, rather than the observed electron emission rate which is constant, beginning instantaneously at the moment the radiation is turned on. Also the wave theory does not explain how the energy of radiation having a wavelength of thousands of atomic diameters may be concentrated in a single atomic electron.

Einstein theorized that the radiation itself consists of discrete quantities of energy, called quanta, which are localized in space and time. Photon is the specific name given to a light quantum. If the radiation is quantized, then the least possible energy of an electromagnetic wave will be $\hbar\omega$. The energy of n photons associated with a particular mode of oscillation in a black body, i.e., the energy of this mode of oscillation, is $n\hbar\omega$. The intensity of electromagnetic radiation is given equally well by either the absolute square of the electric field intensity times the velocity of light, or the number of photons arriving per cm^2-sec times the energy of each photon, $\hbar\omega$. This is very different from Planck's initial interpretation, which suggested that the material oscillators in the walls of the box had quantized energies, rather than the radiation field itself.

Conservation of energy requires that, if no energy is lost due to collisions with atoms in the metal, an electron ejected from a metal surface by the absorption of an electromagnetic wave of frequency ω will have a kinetic energy determined from Einstein's simple equation

$$E = \hbar\omega - W \qquad (3.1)$$

where W is the work function or energy required to free an electron from the attractive force field of the metal. Millikan and others quickly confirmed this relationship between electron energy and radiation frequency experimentally, and observed that no matter how intense the incident radiation, no electrons were released by incident waves of frequency less than $W/\hbar$.

Conversely, Franck and Hertz observed that in a gaseous discharge tube atoms bombarded by an electron beam remain unexcited until the electron kinetic energy is made greater than $\hbar\omega$ for the atomic spectra of the excited atoms. Until excitation, the electron beam is not absorbed nor deflected from the detector, showing that the electrons behave as though in a vacuum and the gas atoms were not there. But as soon as $\hbar\omega$ of a spectral line is exceeded, the electron beam disappears and light of this frequency is produced, showing that electron kinetic energy can be accepted by the atoms only if it produces a quantum jump, and that de-excitation occurs with the creation of a quantum of radiant energy. (The atoms could not be excited by successive smaller nudges, for example.) Therefore, the inter-action of radiation with matter occurs by the loss or gain of discrete amounts of radiation energy called photons. Electromagnetic waves behave in this regard like little lumps or packets of energy.

Problem 3.1 Mercury vapor has two strong absorption lines, at $\lambda = 2537$ Å and $\lambda = 1849$ Å. At what voltages would you expect a current drop in a Franck–Hertz experiment on mercury vapor?

3.2 THE COMPTON EFFECT

A. H. Compton observed experimentally that monochromatic x-rays in-cident on a scattering target are scattered at all angles from the atoms and from the atomic electrons in two frequency components, the component being scattered from the entire atom has the original frequency of the radiation and the other has a slightly lower frequency, decreasing with larger observed scattering angles. He successfully interpreted his result in terms of scattering by electrons in which momentum and energy were conserved.

As a consequence of his electromagnetic theory of light Maxwell showed that a beam of radiation has momentum, equal to the energy trans-ported by the beam divided by the velocity of light. This is well confirmed, for example, by observing the spinning of a radiometer vane, mounted in a vacuum, due to radiation reflected from its arms, or by a more recent example, the orbital changes of the artificial earth satellite Echo I brought about by solar radiation pressure. The minimum altitude of Echo I above the earth decreases and increases periodically by hundreds of kilometers due to solar radiation pressure. Hence, each photon should have a momentum whose magnitude is given by the quantum energy divided by its velocity, $\hbar\omega/c$. If a photon were scattered by an electron in a target material, momentum would be exchanged, and since the electron is essentially stationary, the photon would leave the scattering volume with the speed of light, but with less momentum or energy.

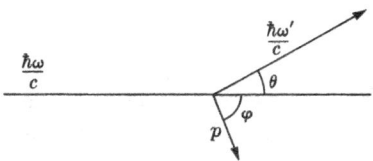

Figure 3.1 *Illustration of a collision between a quantum and an electron initially at rest, showing momenta exchanged.*

If a photon is deflected through an angle θ by collision with an electron at rest as illustrated in Figure 3.1, then conservation of momentum and energy require

$$
\left.
\begin{aligned}
\frac{\hbar\omega}{c} &= \frac{\hbar\omega'}{c}\cos\theta + \frac{mv}{\sqrt{1-(v/c)^2}}\cos\varphi \\[2mm]
\frac{\hbar\omega'}{c}\sin\theta &= \frac{mv}{\sqrt{1-(v/c)^2}}\sin\varphi \\[2mm]
\hbar\omega &= \hbar\omega' + mc^2\left(\frac{1}{\sqrt{1-(v/c)^2}}-1\right)
\end{aligned}
\right\}
\tag{3.2}
$$

three simultaneous equations in the unknowns ω', v, and φ. Solving for ω' we find

$$
\frac{c}{\hbar\omega'} - \frac{c}{\hbar\omega} = \frac{1}{mc}(1-\cos\theta)
$$

or using the wavelength $c/\omega = \lambda$,

$$
\lambda' - \lambda = \frac{\hbar}{mc}(1-\cos\theta)
\tag{3.3}
$$

Thus, in the scattering, the wavelength of the photon (or quantum of light energy) is lengthened while the frequency or energy is lessened. $\hbar/mc$ has the dimensions of length and is called the Compton wavelength for the particle of mass m. The Compton wavelength of the electron is 3.86×10^{-11} cm. The recoil electrons have been observed in cloud chambers, and the momentum and energy of the recoil electrons have been measured and found to agree with Compton's predictions. It is especially important to note here that the precise accounting of the photon frequency and wavelength implicit in Eqs. (3.2) and (3.3) is possible only for long times $(t > |\omega - \omega'|^{-1})$ and over large volumes $(v > |\lambda' - \lambda|^3)$.

Problem 3.2   What is the average pressure on an interplanetary solar sail at the earth's orbit where 1.374×10^6 ergs/cm²-sec of solar radiation power is incident?

Problem 3.3   What is the Compton wavelength of a piece of chalk having a mass of 100 gm? a uranium-238 nucleus? a proton? a π meson, whose mass is 264 electron masses? How does this last value compare with the range of nuclear forces, 1.7×10^{-13} cm?

3.3 RECONCILIATION OF THE PHOTON AND WAVE THEORIES OF LIGHT

The preceding discussion makes it apparent that light frequently exhibits particle-like behavior. That is, it travels and interacts with matter as a small bundle of energy and momentum well localized in space and time. Indeed, in any medium where the absorption and index of refraction change negligibly within a radiation wavelength, the wave nature of light is not apparent and ray optics becomes an excellent approximation. Newton, as is well known, was able to explain many of his observations by a corpuscular theory of light.

Yet when the pertinent dimensions of an obstruction to a photon become sufficiently small, diffraction and interference phenomena are observed which have been successfully interpreted only on the basis of a wave theory (see Fig. 3.2). We here undertake a discussion of interference

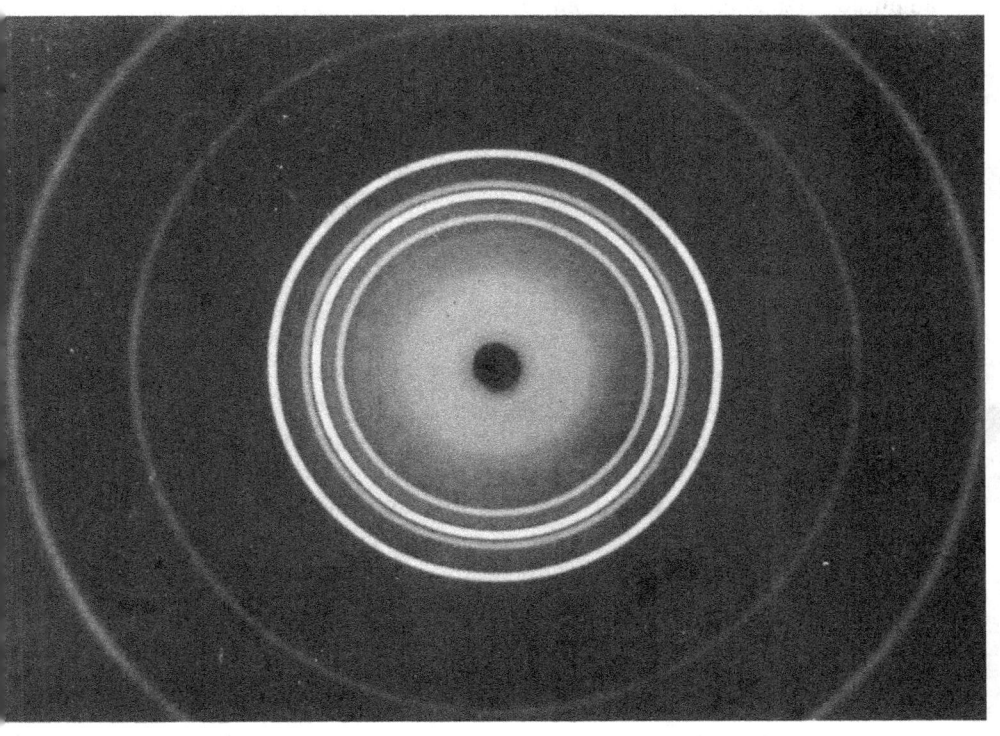

Figure 3.2 *X-ray diffraction pattern of polycrystalline aluminum. (Courtesy of Mrs. M. H. Read, Bell Telephone Laboratories, Murray Hill, New Jersey.)*

in terms of photons whose average position and motion are determined by means of wave packets extending over large volumes of space and long times. In the wave theory of light the light intensity or light energy/cm²-sec, which is necessarily always the observed quantity, is computed by taking the square of the absolute value of a complex quantity called the wave amplitude. (In Maxwell's electromagnetic theory of light the wave amplitude is the electric or magnetic field vector of the wave.) First the phase of any particular part of the total wave disturbance at a given point is computed along the wave path up to the given point. The total wave amplitude at any particular point is then computed by summing or integrating all the contributions from all possible wave paths. The square of the absolute value of this quantity is the intensity. (See the texts on optics listed in the Bibliography of this chapter.)

The frequency spectrum of any particular atomic bright-line radiation (which can be observed, incidentally, only in experiments involving the emission of many photons) shows a bright center with monotonically decreasing intensity as a function of frequency difference with the frequency of the center of the line. The finite frequency breadth of the line causes the Fourier transform of the frequency spectrum to represent (as it must) a sharp pulse in time of very short duration $\Delta t = (\Delta\omega)^{-1}$ where $\Delta\omega$ represents the difference in frequencies at which the intensity is at a maximum and at half maximum value ($\sim 10^{-8}$ sec for ordinary atomic spectra). The spatial extent of the radiation, or rather the length within which there is an appreciable probability of finding a photon, would be the velocity of light times the time duration of the probability of emission (~ 3 meters for atomic spectra in a vacuum). A representative wavelength of the radiation might be 5×10^{-7} m (i.e., about 5000 atomic diameters).

The light intensity may be expressed alternatively as the energy of a photon (whose position is represented by a wave packet) times the number of photons of that frequency arriving per cm²-sec. The average number of photons per cm²-sec or probability of finding one photon per cm²-sec at any particular point is found, as just stated, by computing the amplitude of the light intensity. Suitably normalized, the amplitude of the light intensity (the electric field vector) may be considered as proportional to a *probability amplitude* (which in general is complex), whose absolute value squared gives the probability of counting a photon within, say, the volume of a silver grain in a photographic plate. A diffraction experiment could be performed, in principle, by a source which emitted one photon per minute. A detector placed behind a diffraction grating would then count one photon per minute, each photon being absorbed by a particular detector atom at a particular position. The average counting rate, or total observed counting rate at any position, results from the interference of the probability

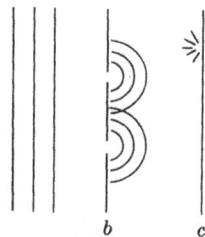

b   *c*

Figure 3.3 *Illustration of a photon being registered at c (indicated by a star) after the plane wave incident on the slits at b is diffracted and expands in concentric circles beyond the slit openings.*

wave of each photon simultaneously transmitted through all the separate openings in the grating. Niels Bohr has given a very nice illustration of this phenomenon which is depicted in Figure 3.3. A more detailed treatment of diffraction of probability waves will be presented in Chapter 4.

3.4 WAVE CHARACTERISTICS OF PARTICLES

Uncertainty Principle   In Chapter 2 we discussed the fundamental relationship between the timing of an electromagnetic signal and its frequency. Equation (2.16) expresses the fact that if we attempt to specify the duration of an electromagnetic wave, we are unable to assign the wave a unique frequency. If a light signal has a finite duration (not infinitesimal), we are unable to know precisely when the light wave is absorbed in a detector. Furthermore, if the light signal has a finite duration (not infinite), the frequencies of the photons which are absorbed are also not precisely known, since the Fourier analyzed frequency components of the signal consists of many frequencies. Rephrased, Eq. (2.16) says that the uncertainty in the timing of a light signal (or change in amplitude of a wave) times the uncertainty in specifying the frequency can never be less than unity. Equation (3.3), expressing the Compton effect in which the scattered light gives an unknown momentum

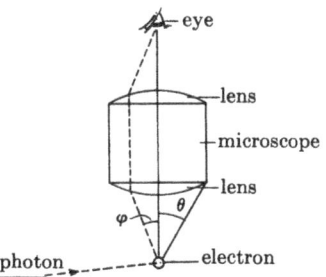

Figure 3.4   *Illustration of path of photon (dashed line) scattered from an electron into a microscope.*

to a particle, enables us to illustrate that that same uncertainty is necessarily with us in any and all measurements on particles. Suppose we attempt to measure the position and velocity of an electron simultaneously with an optical microscope (see Figure 3.4). The position of the electron will be uncertain due to the diffraction of light. A photon will be reflected from the electron and travel through the microscope to our eye but its point of origin, the electron, will be uncertain due to the wave-like diffraction of the photon by the microscope aperture. The total uncertainty in position, or reciprocal of the resolving power of the microscope, is found from elementary physical optics to be $\Delta x = 1.22\lambda/\sin\theta$ where θ is the angle made by the microscope aperture with the axis of the instrument at the electron. Knowing that the momentum of a photon is $h\nu/c = h/\lambda$, we know that the amount of momentum exchanged by the electron and photon is $p\sin\varphi$ where φ is the angle the photon is scattered into by the electron. The maximum angle of scattering is θ, φ is unknown between the limits $\pm \theta$, and therefore the momentum acquired by the electron is equal to or less than $p \sin \theta$. Hence, the product of the uncertainty in position and momentum of the electron parallel to a particular direction, say the x axis, after observation would be

$$\Delta x \Delta p_x = (1.22\lambda/\sin \theta)(h \sin \theta/\lambda) \gtrsim h \tag{3.4}$$

$\Delta x \Delta p_y$, however, can be zero. We may obtain the same relation for photons if Eq. (2.16) is rewritten $[\Delta(h\nu/c)](c\Delta t) \geqslant h$, since $c\Delta t$ is the uncertainty in position of a photon. Many ingenious experiments have been proposed in an attempt to disprove this relation, but none has met with success. This relationship between complementary physical variables, in this case position and momentum, is a well substantiated property of nature. Similarly, the uncertainty in the energy of a particle and in the length of time it has that energy has been shown to satisfy the identical condition Eq. (2.17). Heisenberg enunciated the uncertainty principle and has discussed the intimate relation between it and wave mechanics. By methods beyond the scope of this text, one could state the uncertainty principle as a fundamental postulate and derive quantum mechanics therefrom.

Problem 3.4   What is the approximate expected lifetime of an atomic state whose emission wavelength of ~ 5000 Å can be determined with a precision of one part in 10^3, i.e., $\Delta\lambda/\lambda = 0.001$?

Problem 3.5   If the lifetime of a state is 10^{-9} seconds, what is the line width or uncertainty in the energy of the emitted radiation?

3.5   THE DISCRETENESS OF ATOMIC ENERGY LEVELS

Another characteristic shared by particles and waves confined within a small volume is the discreteness of the energies or frequencies which the system may have. A great many diverse observations, such as the photoelectric or Compton effects or, in general, the emission of electrons by material particles, have well confirmed the fact that atoms contain electrons having mass negligible compared to observed atomic masses, and electromagnetic radiation emitted by atoms has long been identified with the motion of atomic electrons. Zeeman observed that if atoms were placed in a strong magnetic field, the frequency of the emitted radiation was shifted by an amount which is an integral multiple of $eH/2m$ where H is the magnetic field intensity and e/m is the ratio of the charge on the electron to the mass of the electron. $n\hbar eH/2m$ is the magnetic energy, in the applied field H, of a circular electron current. By observing the sign of the shift in frequency and the circular polarization of the radiation parallel to the magnetic field axis, Zeeman and Lorentz determined that the charge, which was the evident source of the radiation, was negative and circulating in the magnetic field. This is known as the *normal* Zeeman effect.

The size of the entire atom is obtained from the density of the material and Avogadro's number; that is, by determining the volume per atom of a macroscopic quantity of a given material. Atomic radii are thus determined to be of the order of 10^{-8} cm. By scattering alpha particles from thin films of metal, however, Rutherford and subsequent experimenters observed unequivocally that virtually all the atomic *mass and positive charge* are concentrated within a radius of the order of 10^{-12} cm. The observed few, large-angle deflections of the alpha particles cannot be interpreted in any other way.

It was natural for physicists of Rutherford's era to attempt to construct a theoretical atomic model resembling the solar system, the analogy resulting from electromagnetic forces acting between orbiting electrons and positive nuclei replacing the gravitational forces acting between the planets and the sun. One of the main difficulties with this theory is that the centripetal acceleration of the electrons would cause them to radiate electromagnetic radiation copiously, which would be observed as a "broad-band" frequency source, not as a single "bright-line" source. Moreover, this theory would predict a rapid energy loss by radiation; thus atoms would not be stable and would lose all their potential energy of charge separation in a small fraction of a second in collapsing to a volume only 10^{-12} cm in radius.

Since atoms are observed to be stable and to emit light of only a few discrete frequencies unique to each atom, one must conclude that atomic electrons can have only certain discrete energies, which suggests that their behavior too is quantized and may be satisfactorily predicted from solutions of a wave equation which satisfy certain boundary conditions resulting from the small confinement of the electrons in the nuclear electrostatic field of force. That is, the potential and kinetic energies of electrons in the atomic electric field are limited somehow to only a few particular values, as are the energies of photons confined in a black body. Semiclassically, one has to conclude that only those electron orbital radii are possible consistent with the energy values theoretically inferred from the observed discrete atomic energy levels. How such limitations on the electron energies could arise was not understood; but the existence of these limitations certainly constitutes wave-like behavior of the electrons.

Problem 3.6 Calculate the classical lifetime of a circularly moving electron in the hydrogen atom on the Rutherford model if the electron is originally at $5 \cdot 10^{-9}$ cm from the nucleus. (Hint: The total energy of a particle in a circular orbit in an attractive central force field with radial dependence $1/r^2$ is equal to minus one-half the absolute value of its potential energy. The power radiated is $\partial E/\partial t = \frac{2}{3}(e^2/c^3)a^2$ in cgs units where a is the particle acceleration, or $\frac{2}{3}(e^2/c^3)(a^2/4\pi\epsilon_0)$ in mks units. The centripetal acceleration may be obtained nonrelativistically by equating $m\omega^2 r = e^2/4\pi\epsilon_0 r^2$.)

3.6 THE DAVISSON–GERMER EXPERIMENT

The old corpuscular theory of light was suggested by observations that light traveled in straight lines in a homogeneous medium just as particles do when moving in any force-free field. This theory was proven to be incomplete by the fundamental observation that light shows interference effects and experiences diffraction. Only wave motion may result in these phenomena; hence diffraction observations permanently established the wave nature of light.

Just as electromagnetic radiation exhibits corpuscular characteristics, so also do material particles appear wave-like in their behavior. The analogous behavior of particles and photons led de Broglie to suggest that associated with particles there might be a wave whose length is given by

$$\lambda = h/p$$

Here p is the particle momentum corresponding to the momentum of a photon in the Compton effect, $p_{(\text{light})}$, which is given by

$$\lambda = c/\nu = h/(h\nu/c) = h/p_{(\text{light})}$$

Here λ is the wavelength of the electromagnetic radiation and ν its frequency in cycles per second.

Niels Bohr determined theoretically the allowed planetary orbits of the electrons by placing appropriate conditions on the solar system atomic model described in the previous section. de Broglie showed that the path lengths over one period of the Bohr orbits of atomic electrons were all integral multiples of h/p, suggesting that electrons in these allowed orbits were in actuality characterized by standing waves.

Davisson and Germer subsequently studied the reflection of 200 ev electrons from a nickel crystal and observed a diffraction pattern for the reflected electrons identical to the Bragg reflection pattern for x-rays from crystals. (See Figure 3.5 and compare with Figure 3.2 for photon diffraction on a different crystal.) The number of reflected electrons at any point could be successfully predicted if one assumed the electrons diffracted like x-radiation with a wavelength equal to the de Broglie wavelength of 200 ev electrons (~ 0.027 Å). Many subsequent experiments on the reflection and transmission of atomic particles, protons, and neutrons through crystals have abundantly confirmed that particles experience wave diffraction and interference. These observations are as fundamental to a wave theory of particles as a Young slit experiment or ordinary diffraction grating experiments are to a wave theory of light. How one may construct a wave theory of particles to correspond in success with the Huygens wave theory of light will be undertaken in the following chapter; the necessity for such a theory is established absolutely by these experiments.

Problem 3.7 What are the de Broglie wave lengths for a 10 ev electron? 1 mev electron? A 10 ev proton? A 100 g rifle bullet moving 3000 feet per sec?

Figure 3.5 *Electron diffraction pattern of polycrystalline tellurium chloride. (Courtesy of RCA Laboratories, Princeton, New Jersey.)*

BIBLIOGRAPHY

Bohr, Niels, *Atomic Physics and Human Knowledge*. New York: John Wiley & Sons, Inc., 1958. A detailed discussion of the uncertainty principle and complementary variables is given in Chapter 4.

There are many excellent books on modern physics, a few of which are:

Weidner, Richard T. and Robert Sells, *Elementary Modern Physics*. Boston: Allyn & Bacon, 1960.

Leighton, Robert B., *Principles of Modern Physics*. New York: McGraw-Hill Book Co., 1959.

Richtmeyer, F. K., E. H. Kennard, and T. Lauritsen, *Introduction to Modern Physics*. New York: McGraw-Hill Book Co., 1955.

Sproull, Robert L., *Modern Physics*. New York: John Wiley & Sons, Inc., 1956.

Slater, J. C., *Modern Physics*. New York: McGraw-Hill Book Co., 1955.

A few of the finest elementary texts in physical optics are:

Rossi, Bruno, *Optics*. (Cited in Bibliography of Chapter 2.)

Jenkins, F. and White, H., *Fundamentals of Optics*. (Cited in Bibliography of Chapter 2.)

Andrews, Charles, *Optics of the Electromagnetic Spectrum*. Englewood Cliffs, N.J.: Prentice-Hall, Inc., 1960.

THE FEYNMAN PATH INTEGRALS

AND THE SCHROEDINGER EQUATION

4

4.1 ELECTRON DIFFRACTION

SINCE THE ORIGINAL Davisson and Germer experiment, electron diffraction has become a very common laboratory phenomenon. Diffraction is peculiar to wave motion; the observation of it is central to our understanding of particle behavior. Figure 4.1 illustrates a simple electron diffraction experiment. A monoergic source of electrons at a emits electrons which pass through two slits at b to be counted on a photographic plate c. The two curves at the right of c represent respectively the actual wave-like result obtained, entailing diffraction and interference effects, and the classically predicted result.

A is a graph illustrating the changes in grain density in the photographic film at c as a function of distance parallel to the plane containing the slit openings. While the electrons arrive at the film one at a time, as could be proved by using a scintillation detector at c, the counting rate or film density at the point c_1 in the figure is as much a function of whether the lower slit is open as of whether the upper slit is open. That is, an electron arriving at c_1, for example, is simultaneously influenced in its motion by *both* slit openings. What is wrong with the classical prediction illustrated by the curve B? It says that the electrons will go through *either* the upper *or* lower slit before being counted at the plate. Those electrons

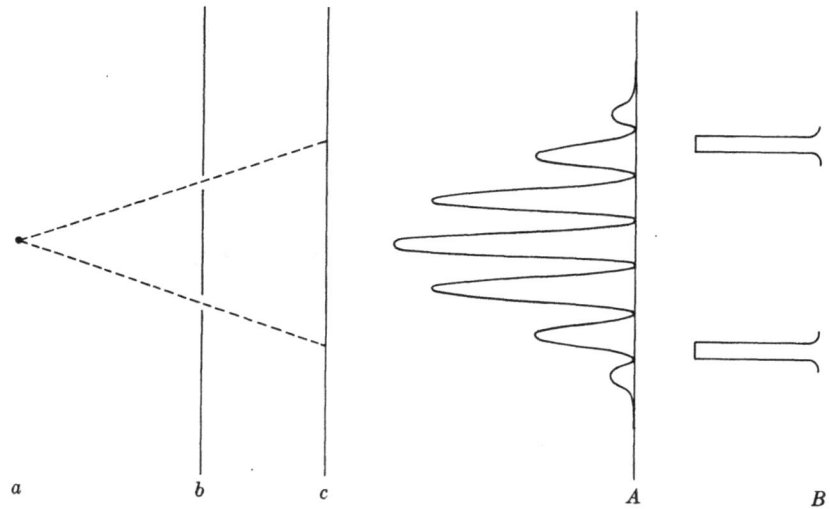

Figure 4.1   *Symbolic illustration of a simple electron diffraction experiment showing monoergic electron source at a, two slits at b, and photographic plate c. A is a graph representing the actual photographic result obtained on c, showing typical wave diffraction and interference maxima and minima. B represents the classically predicted result.*

passed by the upper slit will be expected to contribute to the upper classical counting peak and those passed by the lower slit, to the lower classical peak. The presence of the two peaks would imply that the electrons went through one slit or the other, the electrons in the upper counting peak, for example, having traveled via the upper slit. But if the two slits should be located a distance less than $h/\Delta p$ apart where Δp is the uncertainty in the component of the electron momentum parallel to the slit opening (each slit presumably being less than $h/\Delta p$ wide), then specifying by means of the photographic result which of the two slits the particles went through would mean locating the electrons within a distance less than $h/\Delta p$, in violation of the uncertainty principle. Since we cannot say which of the slits any particular electron

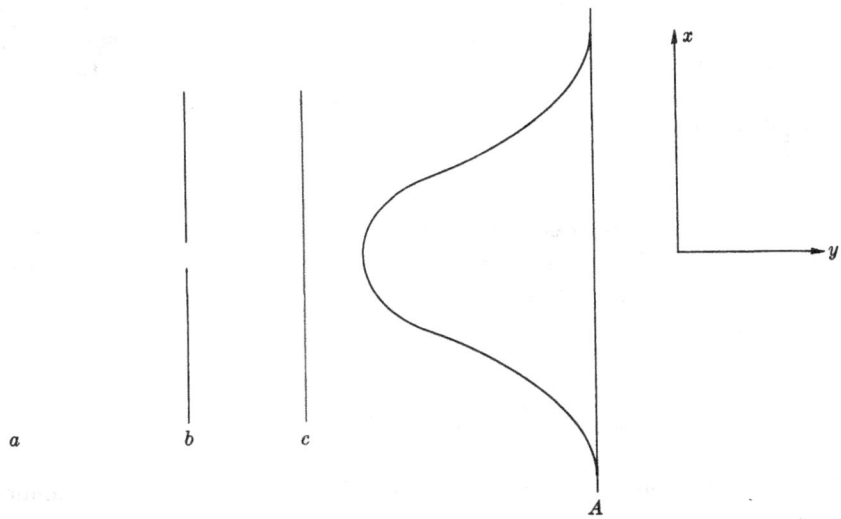

Figure 4.2 *Illustration of a diffraction experiment (a, b, c) and a graph of the counting rate or film density (A) as a function of distance up the screen at c for only one narrow slit open.*

or group of electrons went through, the classical description of an electron going through a particular hole is meaningless.

If the experiment were performed with only one slit open, the curve illustrated in Figure 4.2 would result. A broad, almost uniform distribution (A) results if the slit opening Δx is small. Since the electron penetrates the slit its position at the slit is known with a precision of Δx. Therefore, its momentum in the x direction, Δp_x, can be known to a precision no better than $h/\Delta x$. In traveling a distance $\overline{bc}$ from b to c, therefore, for a time $t = m\overline{bc}/p_y$ it will move in the x direction a distance roughly equal to or less than $\pm \Delta p_x t/m = \pm \hbar \overline{bc}/\Delta x p_y = \pm \lambda(\overline{bc}/\Delta x)$. It is the interference between these broad diffraction distributions from the separate slits which results in the familiar diffraction grating pattern in ordinary physical optics.

A more elegant description of the classical prediction, which has been discussed by Feynman, is the following: The probability of observing an electron at a particular point c emitted from a source at a, P_{ac}, is given by the probability of observing an electron at one of the slits at b, P_{ab}, times the subsequent chance of finding the electron at c, P_{bc}, and then summing over all the slits at b, i.e.,

$$P_{ac} = \sum_b P_{ab} P_{bc} \tag{4.1}$$

As we know, this expression is incorrect since it treats the electrons as though they travel from the source a to a particular opening b and then from this particular opening b to the spot c, and we know it is wrong to state that the electrons travel via any one particular opening.

4.2 DIFFRACTION OF LIGHT

From previous courses the student is quite familiar with a situation similar to the diffraction of electrons, namely, the diffraction of light. *A correct particle theory must predict an electron diffraction pattern identical to that for light.* We will briefly review the theory for the preceding diffraction experiment, using photons instead of electrons.

In this case, the number of photons arriving at a point c_1 on screen c is proportional to the intensity of the light wave, given by the absolute square of the electric field of the wave. The electric field may be expressed as a complex amplitude $e^{i(kx-\omega t)}$ where $k = 2\pi/\lambda = 1/\lambda$ (λ = wavelength). The phase of the wave passed by each slit is given by an integral over the path of the wave. In the case of two slits, the relative number of photons arriving at c_1, or the probability of a photon arriving at c_1, is then given by

$$P_{ac_1} = \left| e^{-i\omega t} \left\{ \exp\left(i \int_{b_1} k \, dx \right) + \exp\left(i \int_{b_2} k \, dx \right) \right\} \right|^2 \tag{4.2}$$

where the two integrals are taken over respective paths through the two slits. If the two paths differ by an odd multiple of $\lambda/2$, then $P_{ac_1} = 0$, a familiar condition for destructive interference. For more than two slits, we have

$$P_{ac_1} = \left| e^{-i\omega} \sum_j \exp\left(i \int_{b_j}^{\cdot} k\,dx\right) \right|^2 \tag{4.3}$$

or in the case of finite slit openings, say in the case of a lens opening or an astronomical mirror, the summation for all the path contributions must be replaced by an areal integration,

$$P_{ac_1} = \left| e^{-i\omega} \int dz \int dy \, \exp\left(i \int_{b(z,y)} k\,dx\right) \right|^2 \tag{4.4}$$

Straight line paths through points in the lens or mirror are assumed for the integrations, since it can be shown that the contribution to the intensity from the summation over the other paths is zero. In general,

$$P_{ac_1} = \left| e^{-i\omega t} \int_{\substack{\text{all} \\ \text{possible} \\ \text{paths}}} \exp\left(i \int_b k\,dx\right) db \right|^2 \tag{4.5}$$

The phase $\int_b k\,dx$ in (4.5) is computed along a particular path b in order to obtain that path's contribution to the total amplitude of the wave originating at a and arriving at c. One must then obtain the total wave amplitude by integrating the contributions from all the separate paths over the lens, mirror, or slit openings. Let

$$\varphi_{ac_1} = \exp\left(i \int_a^{c_1} k\,dx\right), \qquad \varphi_{ab_j} = \exp\left(i \int_a^{b_j} k\,dx\right),$$

and

$$\varphi_{b_j c_1} = \exp\left(i \int_{b_j}^{c_1} k\,dx\right)$$

Since the sum in an exponential argument may be represented as a product of exponentials, Eq. (4.3) may be rewritten

$$P_{ac_1} = |\varphi_{ac_1}|^2 = \left| \sum_j \varphi_{ab_j}\varphi_{b_j c_1} \right|^2 \tag{4.6}$$

which is *not* equal to the classical equivalent for particles which is

$$P_{ac_1} = \sum_j |\varphi_{ab_j}|^2 |\varphi_{b_jc_1}|^2 = \sum_j P_{ab_j}P_{b_jc_1} \tag{4.7}$$

The phase of our complex probability amplitudes φ_{ac_1} for computing the probability of a photon arriving at c_1, is given by the path integral $\int k \, dx$ where k may be better written for our purpose

$$k = \frac{2\pi}{\lambda} = \frac{2\pi\nu}{c} = \frac{2\pi h\nu}{hc} = \frac{p}{\hbar}$$

Similarly,

$$\omega = \frac{E}{\hbar}$$

Thus if we express the physical optics treatment of light diffraction in terms of physical quantities possessed by particles (i.e., energy or momentum instead of frequency or wave length) we arrive at the following description of the diffraction of light:

The probability of a photon arriving to be counted at a particular place c_1 from an origin a is given by the square of the absolute value of the integral, over all contributing paths, of a complex quantity which is a function of the path. The complex quantity of unit modulus has a phase given by $(1/\hbar)\int p \, dx$ where p is the photon momentum and the integral is taken over the path. The complex quantity $\exp\{(i/\hbar)(px - Et)\}$ is a *probability amplitude* or *wave function*, and in customary discussions of diffraction of electromagnetic waves it is given as a factor in the expression for the electric field.

If the polarization of light waves is neglected, the probability of counting a photon on a photographic plate or any other detector is given by the intensity of the wave within the detector volume, which is proportional to $\int |\varphi^2| \, d\tau$, where φ is defined following Eq. (4.5) and $d\tau$ represents a differential element of the detector volume.

4.3 PARTICLE PROBABILITY AMPLITUDES

In constructing a wave theory for the motion of particles too minute to be directly observed in a microscope, we are presented with the great difficulty of having to extrapolate observed macroscopic models of physical behavior to a totally different realm of observation. Our conceptual difficulty is comparable to the problem Huygens might have faced if he had never observed waves in water, strings, or other substances in deriving a wave theory of light. Here we must refrain from applying too closely a physical model or mental picture of submicroscopic particle motion based on macroscopic objects, and refer always only to what are *observed*, those physical consequences of the particle motion which may be detected in laboratory apparatus.

In view of the similar characteristics of light and particles, and particularly since *our objective is to obtain the same diffraction pattern for particles as is obtained for light*, we introduce a theory for particle motion identical to that for electromagnetic wave propagation. That is, we take over complete the methods, functions, and interpretation of the preceding section, substituting particles for photons, with but one modification which concerns the space integral in the argument of the wave function. In optics, a space integral $\int p \, dq$ is taken along the path in order to obtain the phase of the wave function. This integral is well known in advanced classical mechanics where it is given the name "action." A closely related integral over the time, having the same units and called Hamilton's principle function, must be used instead in those physical circumstances where the time-dependent part of the wave function is not independent of the path. Since ω for a light wave is independent of coordinates, and in our discussion of optics we were concerned with equal time intervals for all our paths, $e^{-i\omega t}$ was simply excluded from the integration; but more generally $e^{-iEt/\hbar}$ may be included in the integration, the space integral may be made a time integral by substituting $(dx/dt) \, dt$ for dx, and the entire argument of the wave may be integrated over the time along the wave path. That is,

$$\Psi_{ac} = \exp\left\{\frac{i}{\hbar} \int_{t_a}^{t_c} (p\dot{x} - E) \, dt\right\} \tag{4.8}$$

where t_a and t_c are the times at the ends of the path and p and E are here the momentum and total energy of a particle rather than of a photon as in the previous section. We will use Ψ for particle wave functions instead of φ which we have been using for optical wave functions. If we had to evaluate the integral in (4.8) for a particular example, we would express the momentum, velocity, and total energy of the particle as a function of time and integrate over the time of the particle motion from point a to point c. For nonrelativistic motion in Cartesian coordinates, which we have been using,

$$p\dot{x} - E = 2T - T - V = T - V = L \qquad (4.9)$$

where T is the particle kinetic energy, V the potential energy, and the difference, L, is called the Lagrangian. Therefore, the phase of the probability amplitude for the particle taken over a particular path is given by Hamilton's principle function, the time integral of the Lagrangian for the particle taken along the path of the wave. Hence we may write

$$\Psi_{ac} = \exp\left\{\frac{i}{\hbar} \int_{t_a}^{t_c} (p\dot{x} - E)\, dt\right\} = \exp\left\{\frac{i}{\hbar} \int_{a}^{t_c} Lt\, d\right\} \qquad (4.8a)$$

We may summarize the entire proceedure by quoting Feynman*: "In quantum mechanics the probability of an event which can happen in several different ways is the absolute square of a sum of complex contributions, one from each alternative way. The probability that a particle will be found to have a path $x(t)$ lying somewhere within a region of space-time is the square of a sum of contributions, one from each path in the region. The contribution from a single path is postulated to be an exponential whose (imaginary) phase is the classical action (in units of $\hbar$) for the path in question. The total contribution from all paths reaching x,t from the past is the wave function $\Psi(x,t)$ . . . A probability amplitude is associated with an entire motion of a particle as a function of time, rather than simply with a position of the particle at a particular time."

Thus the probability of finding a particle at c which was initially at a is given by the absolute square of the wave function Ψ_{ac} given in Eq. (4.8), where the integral in (4.8) is evaluated along the classical path of the particle between points a and c.

Equation (4.8) assumes that the particle was known to be at a so that the wave function or probability amplitude of finding the particle at a was something like a delta function. At some time t_c later than t_a the wave function or probability amplitude for finding the particle at any point c

---

* See Bibliography of this chapter.

is given by Eq. (4.8) for Ψ_{ac}. If we wished to know the wave function at some later time t_d in terms of the wave function at c, Eq. (4.6) tells us it would be

$$\Psi_{acd} = \Psi_{ac}\Psi_{cd} = \exp\left\{\frac{i}{\hbar}\int_{t_u}^{t_c}(p\dot{x} - E)\,dt\right\}\exp\left\{\frac{i}{\hbar}\int_{t_c}^{t_d}(px - E)\,dt\right\}$$

$$= \Psi_{ac}\exp\left\{\frac{i}{\hbar}\int_{t_c}^{t_d}(p\dot{x} - E)\,dt\right\} \tag{4.10}$$

Equation (4.10) treats only one of many possible intermediate points c (on the many possible paths) between a and d. If we wish to know the *total* wave function or probability amplitude for finding the particle at d, having known the particle was once at a, we must sum over all intermediate points c as given in Eq. (4.6),

$$\Psi_{ad} \gtrless \sum_c \Psi_{ac}\exp\left\{\frac{i}{\hbar}\int^{t_d}(p\dot{x} - E)\,dt\right\} \tag{4.11}$$

For a particle "diffraction grating," we would sum over the slits c in the grating. For an opening (like an optical lens) we would integrate over the area c of the opening.

The argument in the phase integral of Eqs. (4.10) and (4.11), $p\dot{x} - E$, is the Lagrangian function given in Eq. (4.9), so that Eq. (4.11) may also be written

$$\Psi_{ad} = \sum_c \Psi_{ac}\exp\left\{\frac{i}{\hbar}\int_{t_c}^{t_d}L\,dt\right\} \tag{4.11a}$$

This expression is similar to the expression inside the absolute value signs in Eqs. (4.2–4.5) for the optical wave function. The chief difference between Eqs. (4.10, 4.11) and the optical wave functions in homogeneous media is that $\omega t = Et/\hbar$ is taken outside the integral over space or time in computing the optical wave functions (4.2–4.5).

In general, our initial wave function is not the special case of $\delta(x - a)$ at $t = t_a$, and we will not be concerned with the special example of a particle "diffraction grating" or "lens opening." Hence, in general, we must integrate over the volume elements $dx\,dy\,dz$ rather than the grating or lens surface elements $dy\,dz$. To simplify the discussion and expressions in the following we will assume a one-dimensional space in x and eliminate the y and z coordinates. We then see that if the probability amplitude for finding the particle at position x_1 at time t_1 is $\Psi(x_1,t_1)$, then the probability amplitude for finding the particle at x_2 at time t_2 is given by

$$\Psi(x_2,t_2) = \frac{1}{A} \int \exp\left\{\frac{i}{\hbar} \int_{t_1}^{t_2} (p\dot{x} - E)\,dt\right\} \Psi(x_1,t_1)\,dx_1 \qquad (4.12)$$

$$= \frac{1}{A} \int \exp\left\{\frac{i}{\hbar} \int_{t_1}^{t_2} L\,dt\right\} \Psi(x_1,t_1)\,dx_1$$

which must be integrated over all the values x_1 may assume (i.e., over all one-dimensional x-space). A is a normalization constant, to be determined later, which makes $\Psi(x_2,t_2) = \Psi(x_1,t_1)$ when $x_2 = x_1$ and $t_2 = t_1$.

Problem 4.1   What is the exact expression for the relative number of electrons detected in a photographic plate as a function of distance along the plate in a direction perpendicular to the beam and the crystal axis when a beam of 400 ev electrons bombard a thin crystal of KF N atoms wide, placed perpendicular to the beam, if the plate is also placed perpendicular to the beam one meter beyond the crystal? Assume the crystal cut parallel to the crystal axes. Potassium and fluorine have identical atomic radii of 1.33 Å, and KF forms a simple cubic lattice similar to the NaCl crystal.

Problem 4.2   What is the probability of observing a neutron at an angle of θ radians from the normal to the face of an atomic pile in a plane perpendicular to a long straight crack in the face of the pile, if the neutron is emerging from the crack with energy 0.025 ev and the crack is $2 \cdot 10^{-4}$ m wide? (Hint: In integrating $e^{(i/\hbar)px}$ over the area of the opening ($\int dy$) use the Fraunhofer diffraction formulas approximation; that is, the length x of a path through the point y in the opening is given by $x_0 - y\sin\theta$ where x_0 is the path length of a ray passing through the center of the opening and y is measured from the center of the opening.)

Problem 4.3   Determine where a particle is most likely to be found whose wave function is given by

$$\psi = (1 + ix)/(1 + ix^2)$$

Solution: The probability of finding a particle within a volume V has already been said to be given by the integral of $\int |\psi|^2 \, d\tau$ over the volume. The average value of any property of a state, for example the position x, is defined to be the sum or integral of the product of each possible value of the property times the probability that the state has this value. For example, the probability that a particle is at a position between x and $x + dx$ is $|\psi(x)|^2 \, dx$. The average value of x is found by integrating the expression $x|\psi|^2 \, dx$ over all values of x, and is indicated by $\langle x \rangle$.

Problem 4.4 If $\psi = C/(a^2 + x^2)$, where C is found by requiring $\int |\psi|^2 \, d\tau = 1$, what is the average value of x, $\langle x \rangle$? Of $(x - \langle x \rangle)^2$?

Problem 4.5 A linear harmonic oscillator, for example a mass on a spring, oscillates classically so that the probability of finding the particle at x is given by $P(x) \, dx = C|(\cos \pi r)/L|$, where L is the amplitude of the vibration and C is found by requiring

$$\int_{-L}^{L} P(x) \, dx = 1$$

Calculate the expectation of x, x^2, and $\sqrt{(x - \langle x \rangle)^2}$.

Problem 4.6 Calculate the expectation, that is, average value of x, x^2, and $(\cos \pi x)/a$ for a plane wave whose wave function is $\psi = C(\sin \pi x)/a$ for $0 < x < a$. C is found by requiring that

$$\int_{0}^{a} \psi^* \psi \, dx = 1$$

4.4 THE SCHROEDINGER WAVE EQUATION

In the previous section we attained our objective of formulating a particle theory which would predict a wave diffraction pattern for particles identical with that obtained in the wave theory of light. We have not yet arrived, however, at the stage where we can solve or explain one other confusing breakdown in the classical theory of particles. Specifically, the free particle wave function or probability amplitude we used in the previous section to calculate the interference or diffraction patterns of particles incident on crystals or openings will not help us to predict the bright line spectra from atoms or the experimental consequences of a host of other spherically or cylindrically symmetric, bound state particle problems in which the particles are pretty well confined in regions less than or of the order of their wavelengths. We could, for example, calculate the wave function

$$\exp \frac{i}{\hbar} \int L \, dt$$

for all the possible electron orbits in an atomic potential subject to appropriate boundary conditions, and use this function to predict bright line spectra and other experimental observations, but in many instances it would be an unnecessarily difficult task. The integral approach expressed by Eq. (4.8), being a recent theoretical development, is not the method by which standard problems have been handled, and the reader will find virtually no examples of this method in the literature on the standard problems. The method he will find in the literature, which is frequently more convenient, involves a differential equation instead. In discussing black body radiation in Chapter 1, where we had electromagnetic radiation bound or confined in a box, we were able to derive the correct wave functions for the photons by using Maxwell's electromagnetic wave equation and finding solutions which satisfied the boundary conditions. We could proceed with particle problems in the same way if we had a similar but not necessarily identical wave equation (i.e., a second-order differential equation) for particles. Because of the simplicity of the differential wave equation and its frequent use in the literature, we will now undertake the transformation of (4.12), involving an integration over the paths, to a differential equation, and in later chapters we will stress its use.

If we know the wave function at time t as a function of position x', then the wave function at a slightly later time $t + \epsilon$ as a function of position x will be given by

$$\Psi(x, \ t + \epsilon) = \frac{1}{A} \int e^{(i/\hbar) L \epsilon} \, \Psi(x',t) \, dx' \qquad (4.12a)$$

where the integral must be taken over all the values x' may assume. In

Cartesian coordinates the Lagrangian may be written

$$L = T - V = \tfrac{1}{2}m\left(\frac{x - x'}{\epsilon}\right)^2 - V(x)$$

Substituting L and

$$\xi = x - x', \qquad d\xi = -dx'$$

in (4.12a) and reversing the limits of integration, we obtain

$$\Psi(x, t + \epsilon) = \frac{1}{A} \int e^{im\xi^2/2\hbar\epsilon}\, e^{-i\epsilon V(x)/\hbar}\, \Psi(x - \xi, t)\, d\xi \qquad (4.13)$$

In the limit as ϵ approaches zero, the first factor oscillates rapidly from positive to negative values as a function of ξ for all ξ much different from zero. Therefore only from ξ near zero, where the first factor in the integrand does not oscillate rapidly, do contributions to the integrand for small ϵ occur. That is, only points near x at time t contribute appreciably to the wave function at x at time $t + \epsilon$. Since we are concerned only with ξ and ϵ near zero, $\Psi(x - \xi, t)$ may be expanded in a Taylor's series:

$$\Psi(x, t + \epsilon) = e^{-i\epsilon V(x)/\hbar}\frac{1}{A}\int e^{im\xi^2/2\hbar\epsilon}\left[\Psi(x,t) - \xi\,\frac{\partial\Psi(x,t)}{\partial x}\right.$$

$$\left. + \frac{\xi^2}{2}\,\frac{\partial^2\Psi(x,t)}{\partial x^2} - \dots\right]d\xi \qquad (4.14)$$

Now

$$\int_{-\infty}^{\infty} e^{im\xi^2/2\hbar\epsilon}\, d\xi = \sqrt{2\pi\hbar\epsilon/im}$$

and

$$\int_{-\infty}^{\infty} e^{im\xi^2/2\hbar\epsilon}\, \xi\, d\xi = 0$$

since the integrand is odd, and

$$\int_{-\infty}^{\infty} e^{im\xi^2/2\hbar\epsilon}\, \xi^2\, d\xi = (\hbar\epsilon i/m)\sqrt{2\pi\hbar\epsilon i/m}$$

Expanding the left-hand side of Eq. (4.14) also in a Taylor's series and keeping only the first two terms, we finally obtain

$$\Psi(x,t) + \epsilon \frac{\partial \Psi(x,t)}{\partial t} = \frac{\sqrt{2\pi \hbar \epsilon i/m}}{A} e^{-i\epsilon V(x)/\hbar}$$
$$\times \left[\Psi(x,t) + \frac{\hbar \epsilon i}{m} \frac{\partial^2 \Psi(x,t)}{\partial x^2} + \ldots \right] \tag{4.15}$$

By expanding our wave function $\Psi(x, t + \epsilon)$ in a Taylor's series in powers of ϵ we have obtained an equation in partial differentials. We will now obtain the second-order partial differential wave equation we seek by equating terms multiplied by similar powers of ϵ. Equating the terms which are of zero order in ϵ, we find the normalization constant to be

$$A = \sqrt{2\pi \hbar \epsilon i/m} \tag{4.16}$$

In order to obtain the first-order terms in ϵ we must first expand the exponential containing $V(x)$, and we find that

$$\Psi(x,t) + \epsilon \frac{\partial \Psi}{\partial t} = \left[1 - \frac{i\epsilon}{\hbar} V(x) \right] \left[\Psi(x,t) + \frac{\hbar \epsilon i}{2m} \frac{\partial^2 \Psi}{\partial x^2} \right] \tag{4.17}$$

The zero-orderms ter in ϵ now cancel, and comparing the remaining terms which are first order in ϵ (the ϵ^2 term can be dropped since we care about $\epsilon \ll 1$) and multiplying both sides by $\hbar/i$, we obtain

$$\frac{\hbar}{i} \frac{\partial \Psi}{\partial t} = \frac{\hbar^2}{2m} \frac{\partial^2 \Psi}{\partial x^2} - V(x)\Psi \tag{4.18}$$

which is the differential equation we sought. Equation (4.18) is the *time-dependent Schroedinger equation.*

If the space and time variables are separable, product wave functions may be used. Let

$$\Psi(x,t) = \psi(x)T(t) \tag{4.19}$$

If we substitute Eq. (4.19) into Eq. (4.18) and divide through by $\psi(x)T(t)$ we obtain

$$\frac{\hbar}{i} \frac{1}{T(t)} \frac{\partial T(t)}{\partial t} = \frac{\hbar^2}{2m} \frac{1}{\psi(x)} \frac{\partial^2 \psi(x)}{\partial x^2} - V(x) \tag{4.20}$$

since

$$\frac{1}{\psi(x)} \frac{\partial \psi(x)T(t)}{\partial t} = \frac{\partial T(t)}{\partial t}$$

and

$$\frac{1}{T(t)} \frac{\partial^2 \psi(x)T(t)}{\partial x^2} = \frac{\partial^2 \psi(x)}{\partial x^2}$$

The left-hand side of this equation is independent of x and the right-hand side is independent of t. Since the equation is to be valid for all x and t, both sides must be equal to a constant which is independent of both x and t. Call the constant $-E$; then

$$\frac{\hbar}{i} \frac{1}{T(t)} \frac{\partial T(t)}{\partial t} = -E$$

whose solution is

$$T(t) = e^{-(i/\hbar)Et} \tag{4.21}$$

and the time-independent equation in x, the *time-independent Schroedinger equation*, is

$$-\frac{\hbar^2}{2m} \frac{\partial^2 \psi(x)}{\partial x^2} + V(x)\psi(x) = E\psi(x)$$

or

$$\frac{\partial^2 \psi}{\partial x^2} + \frac{2m}{\hbar^2}[E - V(x)]\psi(x) = 0 \tag{4.22}$$

The extension of the theory to three dimensions is trivial. The result is

$$\nabla^2 \psi + \frac{2m}{\hbar^2}[E - V(\mathbf{r})]\psi = 0 \tag{4.23}$$

for the time-independent Schroedinger equation, and

$$-\frac{\hbar}{i} \frac{\partial \Psi(x,t)}{\partial t} = -\frac{\hbar^2}{2m}\nabla^2 \Psi(x,t) + V(x)\Psi(x,t) \tag{4.24}$$

for the time-dependent Shroedinger equation.

Problem 4.7   Letting

$$\int_{t_0} p\dot{x}\,dt = \int_{x_0}^{x} p\,dx$$

in Eq. (4.8), show that the one-dimensional wave function (4.8) for $\Psi(x,t)$,

$$\Psi(x,t) = \exp\left\{ \frac{i}{\hbar}\int_{x_0}^{x} p\,dx - \frac{i}{\hbar}\int_{t_0}^{t} E\,dt \right\}$$

satisfies Eq. (4.24), if energy is conserved.

4.5 PARTICLE WAVES

There are differences between photons and particles having mass. Just as the nonrelativistic expression for particle kinetic energy, $p^2/2m$, differs from the expression for photon energy, $\hbar\omega$, Eq. (1.5) differs from Eq. (4.24) in the important matter of being second-order instead of first-order in the time derivative. Equations (1.5, 4.24) are, however, both wave equations for the probability amplitudes of photons and particles and, when employed properly, they satisfactorily predict all experimentally observed phenomena to which they have been applied. Whenever the particle position must be defined more closely than permitted by the uncertainty principle, Eq. (4.24) must replace the classical description of particle behavior in which momentum and position are exact and which has proved so useful in analyzing the motion of macroscopic objects. In other words, classical mechanics must be replaced by quantum mechanics whenever critical distances (such as the distance in which particle potential energy changes appreciably, the distance of closest approach of a positively charged particle to an atomic nucleus, or any other distance which the particle must be confined in or excluded from) become less than, or of the order of, the particle wavelength h/mv. The classical equations of motion are still valid, but as we shall see in later chapters their interpretation is different, for Newton's second law, etc., now refer only to the *average* behavior of the particle. Quantum mechanics is the more general and completely correct theory, to which no exception has so far been found. Classical particle motion is a satisfactory approximation to quantum mechanics whenever the particle wavelength is less than any of the relevant dimensions of the problem. In the limit of very short wavelengths or large energies, the classical result must always be obtained. Similar approximations are possible in the theory of light, as is well known. Newton, for example, was able to explain many of his observations on light by means of a corpuscular or ray theory. Similarly, radio engineers are able to analyze the reflection of radio waves in the ionosphere, where the electron density changes slowly, by means of ray optics, since in this example (or for paths of light, e.g., mirages over a hot desert or sunsets) the refractive properties of the medium change little within a wavelength.

Comparison of Eq. (4.23) with Eq. (1.22) leads to the interesting observation that in a constant potential, particle probability waves follow the same path as light waves in a constant or uniform medium, the effective indices of refraction for particles and light being given respectively by

$$n^2 = \frac{E}{E - V}; \qquad n^2 = \frac{\omega^2/c^2}{k^2} = \frac{k_0^2}{k^2} \qquad (4.25)$$

where k_0 is the wave number of the optical radiation in a vacuum, equal to ω/c, and $2mE$ and $2m(E - V)$ are the square of the particle momentum when the potential is zero and different from zero respectively. This comparison of light wave paths to particle paths is not a uniquely quantum mechanical result; indeed Newton's corpuscular theory of light correctly described the path of light. The difference lies in that the index of refraction for a particle theory is the ratio of group velocity in the *medium* to group velocity in the vacuum,

$$n = p_2/p_1 = \sqrt{E/(E - V)};$$

for light waves, on the other hand, n is the ratio of velocity in the vacuum to velocity in the medium.

The entire discussion of wave packets for photons in Chapter 3 applies with the same validity to particle wave packets. The extension of the particle wave packet in space, Δx, for a corresponding spread in momentum Δp, is clearly given by $\Delta x = h/\Delta p$. Identically, a Gaussian wave packet in momentum space with mean square deviation $\Delta p/h$ would, under transformation, yield a Gaussian wave packet in coordinate space with mean square deviation $\Delta x = h/\Delta p$. Instead of Eq. (2.24) we would now have

$$\Psi(x,t) = \frac{1}{2\pi\hbar} \int_{-\infty}^{\infty} B\, e^{-(p-p_0)^2/2(\Delta p)^2}\, e^{(i/\hbar)(px-Et)}\, dp \tag{4.26}$$

where the phase velocity of the wave is now given by E/p. The group velocity is given by $\partial E/\partial p$. Note that the group velocity of the wave packet is equal to the classical velocity of the particle.

Problem 4.8   If the number of remaining excited atoms at t seconds after their excitation at $t = 0$ is $N_0\, e^{-10^8 t}$, what will be the energy spectrum of a measurement of the excited state? Assume the center of the line has an energy of $\hbar\omega_0$.

Problem 4.9   Find the group velocity v_g of a wave packet which describes the motion of a nonrelativistic electron in free space. How does it differ from the velocity of light in a vacuum, c? (That is, is the group velocity for electrons having different momenta, and therefore wavelength, a constant, as it is for light of different wavelengths?)

Problem 4.10   An electron with 2 ev total energy moves in a potential field as follows: $z < 0$, $V = 0$; $z > 0$, $V = 1$ volt. Find the phase and group velocities of the wave packet above and below the xy plane.

Problem 4.11   Find the two lowest energy levels for an electron in a rectangular box measuring $1 \times 2 \times 3$ Å.

BIBLIOGRAPHY

Feynman, Richard P., *Revs. Modern Phys.*, **20**, 367 (1948). This contains the third development of quantum mechanics following some suggestions of Dirac (the first two being Schroedinger's and Heisenberg's original derivations in 1926), upon which this chapter is based.

Bohm, David, *Quantum Theory*. Englewood Cliffs, N.J.: Prentice-Hall, Inc., 1951. See Chapter 3.

USE OF THE WAVE FUNCTION

TO OBTAIN PHYSICAL PROPERTIES

OF A SYSTEM

5

5.1 OBSERVABLES

OUR ENTIRE THEORETICAL KNOWLEDGE of any physical system is now contained in the wave function, and we must find operators which operate on the wave function to give us values for the different properties of the system which we can observe experimentally. Every physical observable (e.g. energy, position, or angular momentum) may be represented by an operator, Q in an eigenvalue equation $Q\psi_n = q_n\psi_n$ whose eigenvalues, q_n, are possible results of a measurement of the physical observable. We have previously encountered an example of the important operator ∇^2 in Chapter 1 and have examined some of its properties. This brings up an important point about operators. They have no meaning unless they operate on something.

If we wish to calculate a particle's position we find it is impossible to obtain a unique value for it in terms of any previous value, because of the uncertainty principle. Classically a later measurement would yield a position given by $x_2 = x_1 + (p/m)t$ but the uncertainty principle requires

an uncertainty in momentum. We may calculate only the *average value*
or the *expected value* for any measurement. If we consider only a one-
dimensional space, for simplicity, the mean value of x or the *expectation of*
x is given by

$$\langle x \rangle = \int \psi^*(x) x \psi(x)\, dx \qquad (5.1)$$

integrated over all space. In this case the order in which we write x and ψ
does not matter, although, as detailed in a later section, in general the order
in which the operator representing a physical observable (position, velocity,
or whatever) and the wave functions are written *does* matter. Thus, the
reader may think of this integrand as $x\psi^*(x)\psi(x)$, that is, the position times
the probability that the particle has that position, in concordance with the
usual definition of mean value in mathematics or classical statistics.

The particle positions at two different times cannot be calculated
precisely since this would require a precise knowledge of the particle's position
and momentum at one time; therefore the particle velocity cannot be
computed directly from values of x. We must use an operator which by
operating on the particle total wave function gives us the velocity or momentum of the particle. (The particle velocity and momentum are linearly
related, as we assume here, *only* in nonrelativistic motion.) If Eq. (4.8) is
written (see Problem 4.7) as

$$\Psi(x,t) = \exp \left\{ \frac{i}{\hbar} \left[\int_{x_a}^{x} p\, dx - \int_{t_u}^{t} E\, dt \right] \right\} \qquad (4.8a)$$

it is clear that the momentum operator is

$$p_x = \frac{\hbar}{i} \frac{\partial}{\partial x} \qquad (5.2)$$

It is also true, in general, when we must sum over all possible paths between
(x_a,t_a) and (x,t).

Similarly, the energy operator is given by

$$E = -\frac{\hbar}{i} \frac{\partial}{\partial t} \qquad (5.3)$$

From this it follows that the Schroedinger time-dependent equation (4.24) may be written as

$$E\Psi = \left(\frac{p^2}{2m} + V(x)\right)\Psi$$

in which p^2 and E are regarded as operators. The operator $p^2 = p_x{}^2 + p_y{}^2 + p_z{}^2$. The sum of two operators A, B is taken: $(A + B)\Psi = A\Psi + B\Psi$. The product of two operators is taken: $AB\Psi = A(B\Psi)$.

In classical mechanics the Hamiltonian is the name given to the sum of the kinetic and potential energy written as a function of position and momentum $H = p^2/2m + V(x)$. The operator expression on the right-hand side of Eq. (4.24) $-(\hbar^2/2m)\nabla^2 + V$, occurs so often that we will frequently refer to this operator on the wave function by the symbol H. Hereafter we will use the letter E to represent the eigenvalue of the H operator. Thus Eqs. (4.23) and (4.24) could be written

$$H\Psi = -\frac{\hbar}{i}\frac{\partial\Psi}{\partial t} = E\Psi \tag{5.4}$$

The second equality is true only if the energy is a constant and E is an eigenvalue of the H operator when operating on Ψ.

The mean value of the momentum, or *expectation* of any measurement of the momentum, may be found by taking the average value or *expectation* of the momentum operator, Eq. (5.2), which is given by

$$\langle p_x \rangle = \int \psi^*(x)p_x\psi(x)\,dx = \frac{\hbar}{i}\int \psi^*(x)\frac{\partial\psi(x)}{\partial x}\,dx \tag{5.5}$$

In general, the expectation of any operator Q we can find to represent a physical observable is given by

$$\langle Q \rangle = \int \psi^* Q\psi\,d\tau \tag{5.6}$$

Note that this is *not* $\int Q\psi^*\psi\,d\tau$ or $\int \psi^*\psi Q\,d\tau$.

If Ψ is an eigenfunction of the operator Q, $Q\Psi = q\Psi$ where q is an eigenvalue of the state or eigenfunction of the operator Q. q is just a number and is a possible result of the measurement of the physical property of the state represented by the operator Q. For example, in Chapter 1 we found that the operator ∇^2 had the eigenvalue ω^2/c^2 when operating on the wave functions of the radiation in a box. The operator ∇^2 represents the reciprocal of the square of the wavelength of the radiation ($\times (2\pi)^2$), and any given value of ω/c which satisfies Eq. (1.11) for integral k, l, and m is a possible result of the measurement of 2π times the inverse of the wavelength of the radiation in the box. A particle plane wave would yield no definite position of the particle; therefore the momentum could be uniquely specified without violating the uncertainty principle. Thus if $\psi = A\, e^{(i/\hbar)p_{0x}x}$, where A is a normalization constant,

$$
\begin{aligned}
\langle p_x \rangle &= \frac{\hbar}{i}\,|A|^2 \int_{-\infty}^{\infty} e^{-(i/\hbar)p_{0x}x}\, \frac{\partial}{\partial x}\, e^{(i/\hbar)p_{0x}x}\, dx \\
&= \frac{\hbar}{i}\,|A|^2 \int_{-\infty}^{\infty} e^{-(i/\hbar)p_{0x}x} \left(\frac{i}{\hbar}\,p_{0x}\right) e^{(i/\hbar)p_{0x}x}\, dx \qquad (5.7) \\
&= p_{0x}
\end{aligned}
$$

The integral in Eq. (5.7) is linearly infinite but identical to $|A|^{-2}$ whereas $\langle x \rangle$, being a definite integral over x, would give no specific well-defined location of the particle, since $\langle x \rangle = (|A|^2/2)[\infty^2 - (-\infty)^2]$. Throughout the later sections of this book whenever we complete a discussion of a new physical observable we will have to find an operator to represent it, so that its expectation may be found by operating with it on the wave function.

A measurement of the relative number of single systems in a particular energy level gives the probability of finding each system in the corresponding group of states. For example, the intensity of light of a particular frequency emitted by a hot (optically thin) vapor, coupled with the transition probability for the atomic emission, would tell us what fraction of the vapor atoms were in the state or group of states having the energy from which the transition was made. If several states have the same energy, the energy level is said to be degenerate. If the level is degenerate, the relative probability of finding the system in any particular state must be found by means of an operator which has a different eigenvalue for each state in the same energy level.

In three dimensions, the momentum operator in coordinate space operating on the wave function is

$$
\mathbf{p}\psi = \frac{\hbar}{i}\,\nabla\psi \qquad (5.8)
$$

The reader is already familiar with the use of alternative representations in either one of a pair of complementary variables in Chapter 2. The momentum–space wave function is defined as the Fourier transform of the coordinate–space wave function. When the wave function is a function of momentum rather than space it is said to be expressed in the momentum representation. If the momentum rather than coordinate space is the independent variable in which the probability amplitude is expressed, then we have, corresponding to Eqs. (5.2, 5.8),

$$x\Phi = -\frac{\hbar}{i}\frac{\partial}{\partial p_x}\Phi$$

or

$$\mathbf{r}\Phi = -\frac{\hbar}{i}\nabla_{\mathbf{p}}\Phi \tag{5.9}$$

where Φ is the wave function expressed in momentum space (the momentum transform of $\Psi(x,t)$) and $\nabla_{\mathbf{p}}$ is the gradient in momentum space, the derivatives being taken with respect to the degrees of freedom in momentum space (e.g., p_x, p_y, p_z). In this connection it might be noted that

$$p_x\varphi(p_x) \neq \frac{\hbar}{i}\frac{\partial \varphi(p_x)}{\partial x}$$

because $\varphi(p_x)$ is a function of p_x only and the right-hand side is meaningless. ($\Phi(\mathbf{p},t)$, $\varphi(\mathbf{p})$ are analogous to $\Psi(\mathbf{r},t)$, $\psi(\mathbf{r})$.)

The probability of finding a particle in a given volume of momentum space is given by

$$\int \varphi^*(\mathbf{p})\varphi(\mathbf{p})\,d\mathbf{p} \tag{5.10}$$

where the region of integration is the given volume.

5.2 COMMUTATION RELATIONS

Commutation consists in reversing the order of two quantities in an algebraic operation. For example, the scalar product of two vectors is the same when the order of the two vectors is reversed. That is, since $\mathbf{A} \cdot \mathbf{B} = \mathbf{B} \cdot \mathbf{A}$, $\mathbf{A}$ is said to commute with $\mathbf{B}$ in this operation. In the vector product of two vectors, on the other hand, the two quantities do not commute:

$$\mathbf{A} \times \mathbf{B} \neq \mathbf{B} \times \mathbf{A} \qquad (5.11)$$

Application of commutation relations between appropriate operators reveals a great deal about the relationships between various physical observables. Commutation relations are of considerable utility and furnish a powerful tool for the further elucidation of our subject.

The average value or expectation value of any operator f is calculated in the coordinate representation from

$$\langle f \rangle = \psi^*(x) f \psi(x) \, d\tau \qquad (5.12)$$

As we have already seen

$$\langle x \rangle = \int \psi^*(x) x \psi(x) \, d\tau$$

$$\langle p_x \rangle = \left\langle \frac{\hbar}{i} \frac{\partial}{\partial x} \right\rangle = \int \psi^*(x) \frac{\hbar}{i} \frac{\partial \psi(x)}{\partial x} d\tau \qquad (5.13)$$

where the operator must be expressed in coordinate space since the state is represented in coordinate space. We can now easily demonstrate that operators may not commute, so that the order in which they are written is important.

$$p_x x \psi = \frac{\hbar}{i} \frac{\partial}{\partial x}(x \psi) = \frac{\hbar}{i}\left(x \frac{\partial \psi}{\partial x} + \psi \right)$$

$$= \left(\frac{\hbar}{i} + x p_x \right) \psi$$

$$\neq x p_x \psi$$

Therefore, a coordinate and its complementary momentum do not commute.

The *commutator* of x and p_x is defined as

$$[x, p_x] = x p_x - p_x x \qquad (5.14)$$

and is a nonzero operator on the wave function, with the expectation

$$\langle [x, p_x] \rangle = \frac{\hbar}{i}\left\{ \int \psi^* x \frac{\partial \psi}{\partial x} d\tau - \int \psi^* \psi \, d\tau - \int \psi^* x \frac{\partial \psi}{\partial x} d\tau \right\}$$

$$= -\frac{\hbar}{i} \qquad (5.15)$$

This is not too surprising, however, since $\langle x\, p_x \rangle$ is a meaningless symbol anyway, because x and p_x cannot be measured simultaneously with infinite precision. (On the other hand, a quantity such as $(\hbar/x) + p_x$ can be measured in principle even if $\hbar/x$ and p_x cannot be measured separately.) *In general, only operators of simultaneously measurable physical observables commute.* x and p_x are not simultaneously measurable, for example, and as we have just seen, they do not commute.

It is possible to know the value of a sum $A + B$ without knowing the separate terms. In fact, it may not be possible to measure both terms simultaneously. For example, the energy may have a precise value even though $p^2/2m$ and $V(x)$ are not separately simultaneously measurable and cannot both have a precise value since they do not commute with one another. The expectation value of the energy is found, from the relation

$$H = \frac{p^2}{2m} + V(\mathbf{r}) = E$$

to be

$$\langle H \rangle = \int \psi^* \left\{ -\frac{\hbar^2}{2m} \nabla^2 \psi + V(\mathbf{r})\psi \right\} d\tau \tag{5.16}$$

We now show how to compute the time derivative of an expectation value.

$$\frac{d\langle A \rangle}{dt} = \frac{d}{dt} \int \Psi^* A \Psi\, d\tau$$

$$= \int \frac{\partial \Psi^*}{\partial t} A \Psi\, d\tau + \int \Psi^* \frac{\partial A}{\partial t} \Psi\, d\tau + \int \Psi^* A \frac{\partial \Psi}{\partial t}\, d\tau$$

and

$$H\Psi = -\frac{\hbar}{i}\frac{\partial \Psi}{\partial t}; \qquad \Psi^* H = -\frac{\hbar}{i}\frac{\partial \Psi^*}{\partial t}$$

where the operator on the complex conjugate wave function is written on the right of the wave function. Therefore, after substituting for partial derivatives,

$$\frac{d\langle A \rangle}{dt} = \frac{i}{\hbar} \int \left\{ \Psi^* H A \Psi - \Psi^* A H \Psi + \frac{\hbar}{i} \Psi^* \frac{\partial A}{\partial t} \Psi \right\} d\tau$$

$$= \frac{i}{\hbar} \langle [H, A] \rangle + \langle \partial A / \partial t \rangle \tag{5.17}$$

If in particular the operator A does not have an explicit time dependence we have for example,

$$\frac{d\langle A \rangle}{dt} = \frac{i}{\hbar} \langle [H, A] \rangle$$

$$\frac{d}{dt}\langle x \rangle = \frac{i}{\hbar}\langle [H, x] \rangle$$

$$= -\frac{i}{\hbar}\frac{\hbar^2}{2m}\int \Psi^*\left[\frac{\partial^2}{\partial x^2}(x\Psi) - x\frac{\partial^2\Psi}{\partial x^2}\right]d\tau$$

$$-\frac{i}{\hbar}\frac{\hbar^2}{2m}\int \Psi^*\left[\frac{\partial^2}{\partial y^2}(x\Psi) + \frac{\partial^2}{\partial z^2}(x\Psi) - x\frac{\partial^2\Psi}{\partial y^2} - x\frac{\partial^2\Psi}{\partial z^2}\right]d\tau$$

$$+\frac{i}{\hbar}\int \Psi^*[V(x,y,z)x - xV(x,y,z)]\Psi\,d\tau \qquad (5.18)$$

Since x commutes with

$$\frac{\partial^2}{\partial y^2}, \quad \frac{\partial^2}{\partial z^2}, \quad \text{and } V(x,y,z)$$

only the first integral yields a nonzero result.
Since

$$\frac{d^2(x\Psi)}{dx^2} = 2\frac{d\Psi}{dx} + x\frac{d^2\Psi}{dx^2} \qquad (5.19)$$

Eq. (5.18) becomes

$$\frac{d}{dt}\langle x \rangle = \frac{\hbar}{2im}\int \left\{2\Psi^*\frac{\partial\Psi}{\partial x} + \Psi^*x\frac{\partial^2\Psi}{\partial x^2} - \Psi^*x\frac{\partial^2\Psi}{\partial x^2}\right\}d\tau$$

$$= \frac{\hbar}{im}\int \Psi^*\frac{\partial\Psi}{\partial x}\,d\tau$$

$$= \frac{1}{m}\langle p_x \rangle \qquad (5.20)$$

a result which is not surprising.

As a second example let us find the time derivative of the expectation of p_x:

$$\frac{d}{dt} \langle p_x \rangle = \frac{i}{\hbar} \langle Hp_x - p_x H \rangle$$

$$= \frac{i}{\hbar} \left\langle \left(\frac{p^2}{2m} + V \right) p_x - p_x \left(\frac{p^2}{2m} + V \right) \right\rangle$$

$$= \frac{i}{\hbar} \langle Vp_x - p_x V \rangle$$

since p_x commutes with p^2. The rest of the commutation operation is given by

$$\frac{d}{dt} \langle p_x \rangle = \int \left(\Psi^* V \frac{\partial \Psi}{\partial x} - \Psi^* \frac{\partial V}{\partial x} \Psi - \Psi^* V \frac{\partial \Psi}{\partial x} \right) d\tau$$

$$= - \left\langle \frac{\partial V}{\partial x} \right\rangle = - \langle (\mathrm{grad}\ V)_x \rangle = \langle F_x \rangle \qquad (5.21)$$

where F_x is the x component of the applied force, the force of gravity, for example, or an electrostatic force. Equation (5.21) is the quantum-mechanical version of Newton's second law.

The equivalence of operating on the term to the left of the operator, and operating on the term to the right, (for example, in Eq. (5.18), it could be shown that the operator $\partial^2/\partial x^2$ has the same expectation when operating on ψ^* as when it operated on $x\psi$), is a generally valid result for all operators representing physical observables, provided the complex conjugate of the operator is taken when the operator operates on the term to the right. Such operators as can be used in this way are called Hermitian operators and they are defined as fulfilling the equality

$$\int \psi_i^*(\mathbf{r})[Q\psi_j(\mathbf{r})]\, d\tau = \int \psi_i(\mathbf{r})[Q\psi_j(\mathbf{r})]^*\, d\tau \qquad (5.22)$$

where ψ, ψ^* are eigenfunctions of the operator denoted Q. *All operators Q which represent physical observables are Hermitian.* We will prove this equality valid for expectation values of all operators representing physical observables. The eigenvalue of the operator Q or Q^* must be a real number since the eigenvalue has a direct physical interpretation. The proof of their being Hermitian consists in first noting that

$$(Q\psi)^* = (q\psi)^* = q\psi^*$$

and

$$Q\psi = q\psi$$

Multiplying on the left both sides of the first of these equalities by ψ and the third equality by ψ^*, subtracting one equation from the other, rearranging and integrating over all space, we find that

$$\int \psi(Q\psi)^* \, d\tau - \int \psi^* Q\psi \, d\tau = q \int \psi\psi^* d\tau - q \int \psi^*\psi \, d\tau = 0$$

Usually we will use the notation $\int (\psi_j^* Q)\psi_i \, d\tau$ for the quantity we have written in (5.22) as $\int \psi_i(\mathbf{r})[Q\psi_j(\mathbf{r})]^* \, d\tau$.

Problem 5.1 Show that $\langle p_x \rangle$ may be found in momentum space from the equation

$$\langle p_x \rangle = \int \varphi^*(\mathbf{p})p_x\varphi(\mathbf{p}) \, dp_x dp_y dp_z$$

where the φ's are the Fourier transforms of the ψ's into momentum space.

Problem 5.2 Show that p_x and p_y commute. Show that x and p_y commute. Show that x and y commute. Show that x and $f(x, y, z)$ commute.

Problem 5.3 Show that

$$[f(p_x, p_y, p_z), x] = \frac{\hbar}{i} \frac{\partial f}{\partial p_x}$$

and

$$[p_x, f(x, y, z)] = \frac{\hbar}{i} \frac{\partial f}{\partial x}$$

(Hint: Use the momentum representation to prove the first relation.)

Problem 5.4 Determine which of the following operators are Hermitian:

$$x, \quad x^2, \quad x^3, \quad \frac{d}{dx}, \quad \frac{d^2}{dx^2}, \quad \frac{d^3}{dx^3}$$

$$i\frac{d}{dx}, \quad x\frac{d}{dx}, \quad \left(x\frac{d}{dx} - \frac{d}{dx}x\right), \quad i\left(x\frac{d}{dx} + \frac{d}{dx}x\right)$$

5.3 PARTICLE CURRENTS

In discussing scattering problems or particle motion it is necessary to find a suitable operator to represent the particle current or flow density. We will next discuss exactly what the particle current density is in terms of the particle wave functions. The complete Schroedinger wave equation is

$$i\hbar\frac{d\Psi}{dt} = -\frac{\hbar^2}{2m}\nabla^2\Psi + V\Psi \tag{5.23}$$

The equation for the complex conjugate wave function is

$$-i\hbar\frac{\partial\Psi^*}{\partial t} = -\frac{\hbar^2}{2m}\nabla^2\Psi^* + V\Psi^* \tag{5.24}$$

Multiplying Eq. (5.23) by Ψ^* and Eq. (5.24) by Ψ and subtracting (5.24) from (5.23), we find

$$i\hbar\left(\Psi^*\frac{\partial\Psi}{\partial t} + \Psi\frac{\partial\Psi^*}{\partial t}\right) + \frac{\hbar^2}{2m}(\Psi^*\nabla^2\Psi - \Psi\nabla^2\Psi^*) = 0$$

or

$$\frac{\partial}{\partial t}(\Psi^*\Psi) + \frac{\hbar}{2im}\nabla(\Psi^*\nabla\Psi - \Psi\nabla\Psi^*) = 0 \tag{5.25}$$

$\Psi^*\Psi$ is the expected particle density. The conservation of particle number is expressed by saying that the time derivative of the particle density ρ plus the divergence of the particle current density $\mathbf{j}$ must add up to zero; that is,

$$\frac{\partial\rho}{\partial t} + \nabla\mathbf{j} = 0 \tag{5.26}$$

Equation (5.26) is known as the equation of continuity. From Eqs. (5.25) and (5.26) we are led to postulate that the average value of particle current density $\mathbf{j}$ is given by

$$\mathbf{j} = \frac{\hbar}{2im}(\Psi^*\nabla\Psi - \Psi\nabla\Psi^*) \tag{52.7}$$

Free particles are represented by the plane wave solutions

$$\Psi = A \exp\left\{ \pm \frac{i}{\hbar}\mathbf{p}\cdot\mathbf{r} - \frac{i}{\hbar}Et \right\}$$

of the Schroedinger equation when the potential energy $V = 0$. By substituting the plane wave solutions into Eq. (5.27), we find that

$$\mathbf{j} = \frac{\hbar}{2im}\left[\left(\frac{i}{\hbar}\mathbf{p}\right) - \left(\frac{i}{\hbar}\mathbf{p}\right)\right]\Psi^*\Psi = \frac{\mathbf{p}}{m}\Psi^*\Psi = \mathbf{v}\Psi^*\Psi \tag{5.28}$$

Thus the particle current density is given by the particle velocity $\mathbf{v}$ times the particle density $\Psi^*\Psi$. If the z direction is taken parallel to the incident beam of particles, the beam is represented by a plane wave e^{ikz}. In case the beam has a finite width it might be represented by a Gaussian wave packet

$$\exp\left(ikz - \frac{x^2 + y^2}{a^2}\right)$$

where a^2 is the mean square deviation of $(x^2 + y^2)^{1/2}$.

Problem 5.5   Find the flux of particles through a sphere of radius a concentric with the origin if the wave function is given by

$$\Psi = \frac{A}{r} \exp\left\{\frac{i(\mathbf{p}\cdot\mathbf{r} - Et)}{\hbar}\right\}$$

5.4   CENTER OF MASS COORDINATES

In atomic physics, the nucleus has essentially infinite mass compared with the electrons and one may ignore the negligible part of an electron's energy which is required to conserve the translational momentum of the complete interacting system. That is, the translational energy and momentum of the atom is negligible compared with the potential and kinetic energy of the electron in the atom. In nuclear physics, where the interacting particles may have equal masses, the translational energy and momentum of the complete system in the laboratory frame of coordinates may not be neglected, and the problem is handled best in a center of mass coordinate frame, in which the translational momentum of the complete system is zero. For two particles of mass m_1 and m_2 having laboratory momenta $\mathbf{p}_1$ and $\mathbf{p}_2$, the total kinetic energy is

$$T = \frac{\mathbf{p}_1{}^2}{2m_1} + \frac{\mathbf{p}_2{}^2}{2m_2} \tag{5.29a}$$

while the potential energy of their interaction is a function of $(\mathbf{r}_1 - \mathbf{r}_2)$; that is,

$$V = f(\mathbf{r}_1 - \mathbf{r}_2) \tag{5.29b}$$

Let

$$\mathbf{P} = \mathbf{p}_1 + \mathbf{p}_2, \qquad \mathbf{p} = \mathbf{p}_1 - \mathbf{p}_2$$

$$m = \frac{m_1 m_2}{m_1 + m_2}, \qquad \mathbf{v} = \mathbf{v}_1 - \mathbf{v}_2 \tag{5.30a}$$

where $\mathbf{v}$ is the relative velocity of one particle with respect to the other and m is the *reduced mass* of the particles. The position of the origin of the center of mass coordinate frame is given by

$$\mathbf{R} = \frac{m_1 \mathbf{r}_1 + m_2 \mathbf{r}_2}{m_1 + m_2} \tag{5.30b}$$

while the separation of the two particles is given by

$$\mathbf{r} = \mathbf{r}_1 - \mathbf{r}_2 \tag{5.30c}$$

By substituting Eqs. (5.30) into Eqs. (5.29), we may represent the Hamiltonian in the new coordinate frame:

$$\begin{aligned}
H &= \frac{P^2}{2(m_1 + m_2)} + \frac{p^2}{2m} + V(\mathbf{r}) \\
&= -\frac{\hbar^2}{2(m_1 + m_2)} \nabla_R{}^2 - \frac{\hbar^2}{2m} \nabla_r{}^2 + V(\mathbf{r})
\end{aligned} \tag{5.31}$$

where ∇_R^2 means that the derivatives in the Laplacian ∇^2 are taken with respect to the coordinates of the point $\mathbf{R}$ and ∇_r^2 means derivatives with respect to $\mathbf{r}$.

Since the variables are separable we may write the wave function as the product of the two wave functions in $\mathbf{R}$ and $\mathbf{r}$ coordinates. That is, we let

$$\Psi(\mathbf{R,r}) = \chi(\mathbf{R})\psi(\mathbf{r}) \tag{5.32}$$

By inserting Eq. (5.32) into the wave equation and then dividing by $\chi(\mathbf{R})\psi(\mathbf{r})$, we find that

$$\frac{1}{\chi(\mathbf{R})\psi(\mathbf{r})}\left(\frac{-\hbar^2}{2(m_1 + m_2)}\right)\nabla_R^2\chi(\mathbf{R})\psi(\mathbf{r})$$

$$-\frac{1}{\chi(\mathbf{R})\psi(\mathbf{r})}\frac{\hbar^2}{2m}\nabla_r^2\chi(\mathbf{R})\psi(\mathbf{r}) + \frac{1}{\chi(\mathbf{R})\psi(\mathbf{r})}V(\mathbf{r})\chi(\mathbf{R})\psi(\mathbf{r})$$

$$= \frac{1}{\chi(\mathbf{R})\psi(\mathbf{r})}E\chi(\mathbf{R})\psi(\mathbf{r})$$

Noting that ∇_R^2 operates only on $\chi(\mathbf{R})$ and ∇_r^2 only on $\psi(\mathbf{r})$, rearranging terms, and making cancellations wherever possible, we obtain

$$-\frac{\hbar^2}{2(m_1 + m_2)}\frac{1}{\chi(\mathbf{R})}\nabla_R^2\chi(\mathbf{R}) = \frac{\hbar^2}{2m}\frac{1}{\psi(\mathbf{r})}\nabla_r^2\psi(\mathbf{r}) - V(\mathbf{r}) + E \tag{5.33}$$

The left-hand side of this equation is a function of $\mathbf{R}$ but not $\mathbf{r}$ while the right-hand side is not a function of $\mathbf{R}$. If this equation is to be valid for any value of $\mathbf{R}$ or $\mathbf{r}$, either side must be equal to a constant. Call it E_{cm}, and let $E_r = E - E_{cm}$. Then

$$\frac{\hbar^2}{2(m_1 + m_2)}\nabla_R^2\chi + E_{cm}\chi = 0 \tag{5.34}$$

is the equation for the center of mass motion, which is the same as that for a free particle having a mass $m_1 + m_2$, and

$$\frac{\hbar^2}{2m}\nabla_r^2\psi + [E_r - V(\mathbf{r})]\psi = 0 \tag{5.35}$$

is the equation dealing with the interaction of the particles and their relative motion. Equation (5.34) is of no further interest and Eq. (5.35) will be our only concern. In all of the following, the center of mass coordinate system is assumed and only the relative motion of the particles will be treated.

BIBLIOGRAPHY

Merzbacher, E., *Quantum Mechanics*. New York: John Wiley & Sons, Inc., 1961. Chapters 4 and 8 especially.

Sherwin, C. W., *Introduction to Quantum Mechanics*. New York: Henry Holt and Co., 1959.

THE CONSEQUENCES OF

ABRUPT CHANGES IN POTENTIAL ENERGY

6

TO EMPHASIZE THE DIFFERENCES between our new description of particle behavior and the older, more familiar classical description, we take up in this chapter the extreme wave-mechanical particle behavior which occurs, just as it does for light, where significant changes (in potential energy) occur within a distance comparable to a wavelength. The simplest potential of this type which we could examine is a so-called square well, that is, a potential which is constant everywhere except at only two points where it changes suddenly from one constant value to another constant value (see Fig. 6.6). Such a potential is extremely useful for many problems in physics, particularly in low-energy nuclear physics where the wavelengths of the particles are much longer than the distance in which large changes occur in the nuclear force field. In such cases the details of the shape of the force field cannot be resolved by means of particles whose wavelength is too long to distinguish any short-range potential, and the square well will frequently yield answers just as correct as, in fact indistinguishable from, answers obtained with more complicated shapes.

Geometrical optics, that is, corpuscular or ray optics, yields excellent approximations for many studies of electromagnetic radiation. For example, light waves passing through the atmosphere, or radio waves traveling through slowly varying changes in index of refraction caused by the slowly varying electron concentration in the ionosphere, follow ray-like paths. Geometrical optics is incomplete, however, when applied to such sharp changes in the index of refraction as occur when light is transmitted through a glass lens or a drop of water. It is well known that the path of the ray may be correctly described by ray optics, as Newton first showed, but the reflection of a fraction of the light by a glass-air surface can be predicted only by wave theory. Coated lenses, for example, are a practical application of wave or physical optics.

Force is deduced from the gradient of a potential, and we note that a square well will lead to an infinite force at one point and no force at all in the rest of space. (Such abrupt changes in potential energy do not occur in macroscopic force fields.) When the total energy is conserved, an infinite force resulting from an abrupt change in potential energy causes an abrupt change in kinetic energy and thus an abrupt change in momentum for particles or in wave number for waves. A similar abrupt change in kinetic energy caused by an infinite force acting for an infinitesimal distance would result if a baseball were thrown through a window. (The analogy to a particle experiencing a sudden change in potential energy is incomplete in that the potential energy of the ball is unchanged by penetrating the window, but the kinetic energy in both instances is abruptly lessened as a result of an infinite opposing force acting for an infinitesimal distance.) A similar sudden decrease in wave number occurs when light penetrates into glass, water, or some other solid medium having an index of refraction different from air. In the absence of a gravitational or other force field the path of the baseball thrown through the window could be shown classically to be the same as the path of the light ray in the parallel circumstance of a sudden change in refractive index. The wave-theoretical difference (from ray optics) is that even though the "breaking strength" of the glass may be exceeded, the baseball will have a finite (although minute) probability of being reflected.

6.1 PARTICLES INCIDENT ON A SQUARE BARRIER

Example 1   Reflection of a particle at a square barrier infinite in extent (see Fig. 6.1). The barrier may be represented by

$$x < 0: \quad V = 0$$
$$x > 0: \quad V = V_0$$
(6.1)

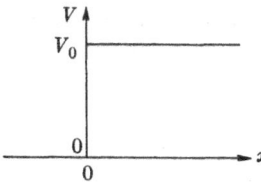

Figure 6.1   *Illustration of a square one-dimensional barrier of infinite extent.*

We further assume the particle energy E to be $> V_0$ and that the particle is incident on the barrier from the left. The potential energies are constant in the two regions on either side of $x = 0$ so that Eq. (4.8) immediately leads to simple plane wave solutions. Since E is a constant we may neglect the time-dependent part of the wave, $e^{-(i/\hbar)Et}$. Letting the particle momenta on the two sides of the barrier be

$$p_1 = \sqrt{2mE}$$

and

$$p_2 = \sqrt{2m(E - V_0)}$$

we find the solutions

$$x < 0: \quad \psi = A\, e^{(i/\hbar)p_1 x} + B\, e^{-(i/\hbar)p_1 x}$$
$$x > 0: \quad \psi = C\, e^{(i/\hbar)p_2 x}$$
(6.2)

where $e^{(i/\hbar)p_1 x}$ represents the incident wave, $e^{-(i/\hbar)p_1 x}$ the reflected wave, and $e^{(i/\hbar)p_2 x}$ the transmitted wave. Equations (6.2) are presented, in customary treatments of this problem, as solutions of the time-independent Schroedinger wave equation, Eq. (4.22), which is

$$\frac{\partial^2 \psi}{\partial x^2} + \frac{2m}{\hbar^2}(E - V)\psi = 0 \tag{6.3}$$

and since the total probability is never greater than one and a particle's energy must be finite, Eq. (6.3) may be satisfied only if the second derivative of the wave function is finite everywhere. Hence a further boundary condition is obtained by requiring that both the wave function and its first derivative be everywhere continuous. Thus at $x = 0$, this boundary condition by use of Eq. (6.2) yields

$$A + B = C \tag{6.4a}$$

and

$$(i/\hbar)p_2 C = (i/\hbar)p_1 A - (i/\hbar)p_1 B \tag{6.4b}$$

whence

$$B = \frac{p_1 - p_2}{p_1 + p_2} A \quad \text{and} \quad C = \frac{2p_1}{p_1 + p_2} A \tag{6.5}$$

The reflection and transmission are obtained from the currents to the left and right of $x = 0$. These are given by the particle velocity in the medium, p/m, times the relative density of particles or the probability per unit volume of finding the particle to the left or right of the potential change. Since the density of particles moving to the left per unit volume is proportional to $|B\,e^{-(i/\hbar)p_1 x}|^2 = B^2$,

$$
\begin{aligned}
\frac{\text{Reflected current}}{\text{Incident current}} &= \left|\frac{B}{A}\right|^2 \frac{p_1}{p_1} = \left(\frac{p_1 - p_2}{p_1 + p_2}\right)^2 \\
\frac{\text{Transmitted current}}{\text{Incident current}} &= \left|\frac{C}{A}\right|^2 \frac{p_2}{p_1} = \frac{4p_1 p_2}{(p_1 + p_2)^2}
\end{aligned}
\tag{6.6}
$$

Note that if there is no potential change, $p_2 = p_1$, and no reflection takes place; that in this case $E = V_0$, $p_2 = 0$, and the particle is 100% reflected; that particle probability current (cf. Section 5.3) is conserved, for the sum of Eqs. (6.6) is unity; and that Eqs. (6.6) exactly correspond to formulas for the same quantities when electromagnetic radiation is perpendicularly incident on a sharp boundary between media of different indices of refraction. In this case the reflection, for example, is given by

$$R = (n_2 - n_1)^2/(n_1 + n_2)^2$$

where n_2/n_1 is the ratio of the velocity of light in medium one to medium two just as p_1/p_2 is the same ratio for particles.

If the particle energy E is less than V_0 we obtain different expressions on the right of the barrier. The particle momentum $\sqrt{2m(E - V_0)}$ on the right of the barrier is imaginary. Letting $q_2 = \sqrt{2m(V_0 - E)}$, we find the solution of Eq. (4.8) to the right of the barrier to be

$$x > 0: \qquad \psi = C\,e^{-(q_2/\hbar)x} \qquad (6.7)$$

That is, the exponent is real and no longer represents an oscillating wave function. (Note that $e^{+q_2\hbar/x}$ is also a solution of Eq. (6.3) in customary solutions of this problem which use Eq. (6.3), but since ψ must be finite everywhere and Eq. (6.7) must hold at $x = +\infty$, this added solution must have a zero coefficient.) Fitting solutions (6.2) for negative x and (6.7) for positive x as we did in obtaining Eq. (6.5), we find in place of (6.5) that for $E < V_0$

$$B = \frac{q_2 - ip_1}{q_2 + ip_1}A$$

$$C = \frac{2ip_1}{q_2 + ip_1}A$$

$$(6.5a)$$

and

$$\text{Reflection} = \left|\frac{B}{A}\right|^2 = 1$$

$$\text{Transmission} = \left|\frac{C}{A}\right|^2 \quad C = 4\frac{p_1 q_2}{(p_1 + q_2)}\,e^{-(2q_2/\hbar)x}$$

$$(6.6a)$$

Since the transmission is zero at $x = \infty$ and no particle absorption has been included in the problem, 100% reflection is also required by particle conservation. Although there is no transmitted current, note that there is a finite probability of finding the particle on the right of the barrier, where classically it cannot be.

Example 2 Reflection and transmission through a finite square barrier depicted in Figure 6.2.

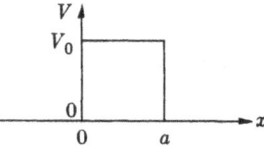

Figure 6.2 *Illustration of a square one-dimensional barrier of limited* extent.

In this case the potential is given by

$$x < 0: \quad V = 0$$
$$0 < x < a: \quad V = V_0 \qquad (6.8)$$
$$a < x: \quad V = 0$$

Again we first assume $E > V_0$ and that the wave is incident from the left, and find the following wave solutions in the three regions representing incident and reflected waves in the left and center regions and only a transmitted wave in the right region:

$$x < 0: \quad \psi = A\, e^{(i/\hbar)p_1 x} + B\, e^{-(i/\hbar)p_1 x}$$
$$0 < x < a: \quad \psi = C\, e^{(i/\hbar)p_2 x} + D\, e^{-(i/\hbar)p_2 x} \qquad (6.9)$$
$$a < x: \quad \psi = E\, e^{(i/\hbar)p_1 x}$$

The reflected amplitude B (made up of a wave reflected at both $x = 0$ and a) and the transmitted amplitude E may be found by requiring again that the wave functions and their first derivatives be continuous at $x = 0$ and $x = a$. The solution requires considerable tedious algebraic manipulation, with the result that

$$\text{Reflection} \ = \ \left|\frac{B}{A}\right|^2 = \left[1 + \frac{4p_1^2 p_2^2}{(p_1^2 - p_2^2)^2} \csc^2 \frac{p_2 a}{\hbar}\right]^{-1}$$

$$\text{Transmission} \ = \ \left|\frac{E}{A}\right|^2 = \left[1 + \frac{(p_1^2 - p_2^2)^2}{4p_1^2 p_2^2} \sin^2 \frac{p_2 a}{\hbar}\right]^{-1}$$

(6.10)

Note that if there is no barrier and $p_2 = p_1$, there is 100% transmission. There is the further very interesting circumstance that if a, the barrier width, is equal to an integral number of half wavelengths, 100% transmission occurs, analogous to the behavior of coated lenses in optics. Otherwise we have the nonclassical result of partial reflection at the barrier.

If $E < V_0$, then classically the particle does not have sufficient energy to jump the barrier, and is 100% reflected. In this case the second of Eqs. (6.9), for $0 < x < a$, no longer represents the correct solution for that region. The correct solution when $E < V_0$ is

$$0 < x < a: \quad \psi = C e^{(q_2/\hbar)x} + D e^{-(q_2/\hbar)x} \tag{6.9a}$$

where $q_2 = \sqrt{2m(V_0 - E)}$. Equation (6.9a) represents exponentially increasing and decreasing solutions within the barrier. Solving the simultaneous algebraic equations resulting from requiring that the wave function and its first derivatives be continuous at both boundaries, we find that

$$\text{Reflection} \ = \ \left|\frac{B}{A}\right|^2 = \left[1 + \frac{4p_1^2 q_2^2}{(p_1^2 + q_2^2)^2} \operatorname{csch}^2 \frac{q_2 a}{\hbar}\right]^{-1} \tag{6.11}$$

$$\text{Transmission} = \left|\frac{E}{A}\right|^2 = \left[1 + \frac{(p_1^2 + q_2^2)^2}{4p_1^2 q_2^2} \sinh^2 \frac{q_2 a}{\hbar}\right]^{-1} \tag{6.12}$$

If $a \gg \hbar/q_2$, then the transmission coefficient is asymptotically given by

$$\left[\frac{q_2}{p_1} + \frac{p_1}{q_2}\right]^{-2} 16 \exp\left(-2\frac{q_2 a}{\hbar}\right)$$

Note that the transmission is nonzero if a is not very many wavelengths thick. Gamow has made an amusing illustration of this situation to a billiard ball which will not remain on the top of the billiard table but will eventually penetrate the sides of the table and drop onto the floor. Figure 6.3 illustrates the wave behavior at the barrier for a roughly comparable to λ and E less than but comparable to V_0.

The particle problem in which $V_0 > E$ has an optical analog in an imaginary index of refraction such as can occur for radio waves in the

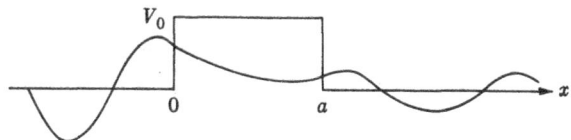

Figure 6.3   *Illustration of the wave behavior at a finite barrier for barrier width roughly comparable to particle wavelength and particle kinetic energy less than but comparable to the barrier height. The particle has insufficient energy, classically, to surmount the barrier. The sum of the incident and reflected waves is shown to the left of $x = 0$; the transmitted wave is illustrated to the right of $x = a$.*

ionosphere or for light waves incident at greater than the critical angle in an optically dense medium. Prism surfaces in binoculars, for example, are arranged so that light is incident on the reflecting surface at greater than the critical angle for glass to air transmission, so that 100% internal reflection will occur for light incident on the air back of the prism.

If an optically flat surface were placed close to the optically flat reflecting surface of the prism, however, some of the light would be transmitted, as could be readily calculated by a formula analogous to Eq. (6.12). This was the principle of an ingenious device composed of two prisms by which, during the Second World War, German scientists attempted to audiomodulate a light beam. One of the prisms in Figure 6.4 was attached to a solenoid activated by an audiofrequency current signal so that the space between the two prisms changed at audiofrequency as the prism was made to vibrate. When the prisms were close together and the separation distance a was small, the system transmitted light to the right across the barrier presented by the air space of thickness a; conversely, when a was large light was reflected from the barrier presented by the air space. Unfortunately the prism could not be made to vibrate well at audio-frequencies.

Equations (6.11) and (6.12) have a useful application in the estimation of radioactive nuclei lifetimes from alpha particle decay. Alpha particles

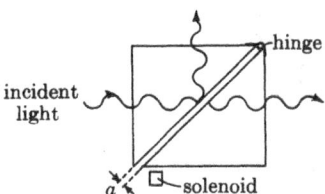

Figure 6.4   *Illustration of the utilization of the optical analog of penetration through a square barrier in quantum mechanics. The transmitted light is modulated in intensity by varying the width (barrier thickness) of the air space between the prisms.*

within radioactive nuclei have a negative binding energy, so that when the alpha particle is separated from the nucleus several Mev of negative binding energy are released. Outside the notably short range of attractive nuclear forces, however, the alpha particle's electrostatic potential energy, due to repulsion by the Coulomb field of the nucleus, far exceeds its total energy. This situation corresponds to a ball on a billiard table which is able to release energy in falling to the floor but does not do so, classically, since it has insufficient energy to surmount the barrier surrounding the table. Quantum mechanically, particles have a finite probability of penetrating finite barriers according to Eq. (6.12) and we should expect some alpha particles to penetrate the barrier illustrated in Figure 6.5.

An alpha particle with a kinetic energy of a few Mev within the nucleus has a velocity of roughly 10^9 cm/sec and strikes the nuclear wall on the order of 10^{21} times per second, since heavy nuclei have a radius of about 10^{-12} cm. Equation (6.12) gives the probability of transmission at each encounter of a particle with the nuclear surface. The thickness of the barrier is energy-dependent and is very roughly 2×10^{-12} cm; the height of the barrier is of the order of 20 Mev so that for an alpha particle having a negative binding energy of 5 Mev (i.e., an energy of 5 Mev greater than the potential energy at infinity), by use of Eq. (6.12) the probability of transmission is

$$T \sim 16 \exp\left(-\frac{4 \cdot 10^{-12}}{10^{-27}}\sqrt{2 \cdot 6.7 \cdot 10^{-24}(20 - 5) \cdot 1.6 \cdot 10^{-6}}\right)$$

$$\sim 16 \exp(-72)$$

$$\sim 10^{-30} \tag{6.13}$$

The probability of decay per second would be

$$10^{21} \times 10^{-30} = 10^{-9}/\text{sec}$$

which corresponds roughly to a decay time of 10^9 seconds or 30 years. That the answer at all resembles the lifetimes of actual nuclei is fortuitous

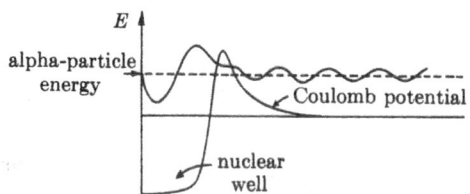

Figure 6.5   *Illustration of total energy (– –) and potential energy (———) in alpha particle radioactive decay. The wave function is also plotted as a solid line.*

in view of the enormous crudities in estimating the exponent from Eq. (6.12), but the procedure at least serves to illustrate the essential features of alpha decay. A more precise calculation (first made by Gamow and Condon) requires the use of the Coulomb barrier rather than a square barrier, and we will make a better estimate in the next chapter.

Problem 6.1 Compute the barrier penetration for electrons having a kinetic energy of 4 ev for a rectangular (one-dimensional) barrier 10 Å wide and 5 ev high. (This problem is analogous to electron tunnelling in the reverse direction in a crystal rectifier.)

6.2 UNBOUND AND BOUND PARTICLES IN A SQUARE WELL

For this case, as illustrated in Figure 6.6,

$$x < 0 \qquad V = 0$$
$$0 < x < a \qquad V = -V_0$$
$$a < x \qquad V = 0$$

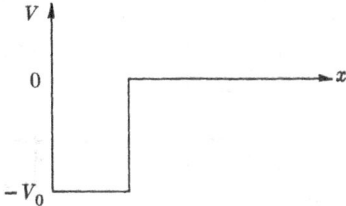

Figure 6.6 *Illustration of square well potential.*

We might examine first the case of $E > 0$. This case is similar to the case of a square barrier with $E > V_0$ except that here $p_2 = \sqrt{2m(E + V_0)}$. Again we should and do obtain transmission resonances at wavelengths equal to twice the well thickness divided by an integer, or, expressed alternatively in terms of the wavelength $(\hbar/mv)$, we obtain transmission resonances for wells which are an integral number of half wavelengths thick. In this case we do not have to go to optical applications such as coated lenses or a Fabry–Perot interferometer for illustration, for particle physics furnishes us with an example which is another signal triumph of the wave theory for particles. Before the development of quantum mechanics, Ramsauer observed experimentally that in the scattering of hot electrons from gas atoms, transmissions or zero-scattering resonances occurred for some atoms at particular electron energies. Instead of increasing with decreasing electron energy as expected, the total collision cross sections of the gas atoms decreased remarkably for the atoms of some elements, as illustrated in Figure 6.7. Quantum mechanics cleared up the mystery of this phenomenon. Basically, a careful, three-dimensional calculation using a screened

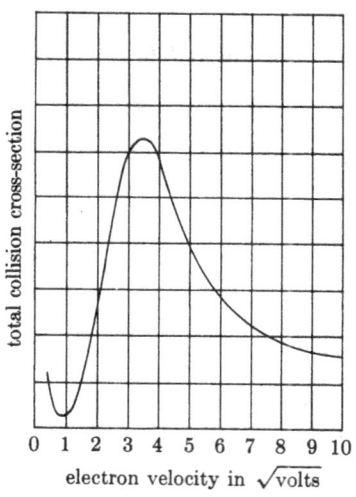

Figure 6.7   *Graph of total collision cross section of gas atoms to incident electrons having energies up to 100 ev.*

Coulomb field can be interpreted as predicting transmission resonances whenever the effective diameter of atoms is an integral number of electron wavelengths (the wavelengths to be taken while the electron is in the screened Coulomb potential).

For the case $E < 0$, $q_1 = q_3 = \sqrt{-2mE} = \sqrt{2mW}$ where $E = -W$, and the wave functions to the left and right of the well vary exponentially. Only inside the well is a sinusoidal wave possible. Therefore, for

$$0 < x < a: \quad \psi = B\,e^{(i/\hbar)p_2 x} + C\,e^{-(i/\hbar)p_2 x} \tag{6.14}$$

where $p_2 = \sqrt{2m(V_0 + E)} = \sqrt{2m(V_0 - W)}$. Since ψ is finite at $\pm\infty$, for

$$x < 0: \quad \psi = A\, e^{(q_1/\hbar)x}$$
$$a < x: \quad \psi = D\, e^{-(q_1/\hbar)x} \tag{6.15}$$

representing exponentially decaying functions to the left and right. In this case there is no transmission out of the well, and we are not interested in the values of the coefficients which represent a standing wave within the well. We are interested, however, in finding any further conditions on our solutions necessary to make them acceptable solutions satisfying the usual boundary conditions. Requiring continuity of functions and their first derivatives at the boundaries, we find

$$A = B + C; \quad A = i\frac{p_2}{q_1}(B - C) \tag{6.16}$$

$$D\, e^{-(q_1 a/\hbar)} = B\, e^{(i/\hbar)p_2 a} + C\, e^{-(i/\hbar)p_2 a} \tag{6.17}$$

$$D\, e^{-(q_1 a/\hbar)} = i\frac{p_2}{q_1}(B\, e^{(i/\hbar)p_2 a} - C\, e^{-(i/\hbar)p_2 a}) \tag{6.18}$$

$$B = +\frac{p_2 - iq_1}{2p_2}A; \quad C = \frac{p_2 + iq_1}{2p_2}A \tag{6.19}$$

Dividing (6.18) by (6.17) and substituting (6.19) in the result, we find that

$$1 = -i\frac{p_2}{q_1}\frac{(p_2 - iq_1)\, e^{(i/\hbar)p_2 a} - (p_2 + iq_1)\, e^{-(i/\hbar)p_2 a}}{(p_2 - iq_1)\, e^{(i/\hbar)p_2 a} + (p_2 + iq_1)\, e^{-(i/\hbar)p_2 a}} \tag{6.20}$$

which with the indicated algebraic manipulation becomes

$$\frac{q_1}{p_2} = \frac{p_2 \sin(p_2/\hbar)a - q_1 \cos(p_2/\hbar)a}{q_1 \sin(p_2/\hbar)a + p_2 \cos(p_2/\hbar)a} = \frac{\tan(p_2/\hbar)a - (q_1/p_2)}{1 + (q_1/p_2)\tan(p_2/\hbar)a} \tag{6.21}$$

Let $\varphi = \tan^{-1}(q_1/p_2)$, then Eq. (6.21) becomes

$$\tan\varphi = \tan\left(\frac{p_2}{\hbar}a - \varphi\right) \tag{6.22}$$

This equation can be satisfied only when $\varphi = (p_2/\hbar)a - \varphi + n\pi$ where n

is an integer. That is,

$$\varphi = \frac{p_2 a}{2\hbar} + \frac{n\pi}{2} \tag{6.23}$$

Substituting Eq. (6.23) into the definition of φ and using the definitions of p_1 and p_2 in terms of W and V_0, we obtain the necessary relation between E, V_0, and a in order for the boundary conditions to be satisfied:

$$\text{for even } n: \quad \sqrt{\frac{W}{V_0 - W}} = \tan \sqrt{2m(V_0 - W)} \, \frac{a}{2\hbar} \tag{6.24a}$$

$$\text{for odd } n: \quad \sqrt{\frac{W}{V_0 - W}} = -\cot \sqrt{2m(V_0 - W)} \, \frac{a}{2\hbar} \tag{6.24b}$$

Equation (6.24) illustrates the vital quantum mechanical result that *the energy of a bound particle can have only certain discrete values*. The result is similar to the familiar condition in wave motion for standing waves. For example, if a wave in a string is constrained between two fixed points, standing waves of discrete frequencies occur in the string. The reader may readily surmise that this result contains the essence of the explanation for the discrete energies possessed by electrons in atoms, as revealed by the bright-line spectra they radiate. Every bound particle can exist only in discrete energy states, similar to the constraints Planck discovered for the radiation waves in a black body. In the case of the massive planets bound to the sun, these different energy levels are so close together in value as to be totally undiscernible to any experimenter, and classical theory, which would predict a continuum for planetary motion, can thus certainly be thought of as exact. If h were much larger than it is, however, and classical physics thus did not apply, collision with another astronomical body would result in no disturbance of the earth's motion unless enough energy was exchanged for the earth to make a quantum jump to another energy state. For sub-microscopic systems, the prediction that no collisional excitation takes place unless the energy transferred is greater than some minimum value is well brought out by the Frank and Hertz experiments on electron beams in a gaseous discharge tube, discussed in Section 3.1.

Equation (6.24) results because the derivative as well as the wave function must be continuous at the boundaries of the well. Only for unique values of the energy (or the wavelength) can the wave functions inside be smoothly joined to the decaying exponential functions outside the well. Figure 6.8 illustrates the three lowest energy wave functions inside the well which satisfy the boundary conditions.

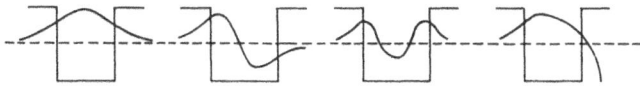

Figure 6.8 *Illustration of the three lowest energy wave functions possible in a square well potential and a fourth impossible wave function requiring a physically unrealizable exponentially increasing wave function outside the well on the right for continuity at the well boundary.*

We can apply Eq. (6.24) to find the energy binding the neutron and the proton in a deuteron. As mentioned earlier, at low energies as in this case the shape of the potential describing nuclear interactions is indistinguishable from a square well. The deuteron interaction occurs in three dimensions and thus will be treated in spherical coordinates.

The deuteron ground state is a spherically symmetric state. A spherically symmetric standing wave depends only on the radial coordinate and is given by the spherical form of Eq. (4.8). This is analogous to a spherically expanding wave in physical optics emanating from a pinhole.

$$r < a: \quad \psi = A_1[e^{(i/\hbar)p_2 r}/r] + A_2[e^{-(i/\hbar)p_2 r}/r]$$
$$r > a: \quad \psi = Be^{-q_1/\hbar}/r \tag{6.25}$$

where a is the range of nuclear forces between neutron and proton in the deuteron, $p_2 = \sqrt{M(V_0 - W)}$, and $q_1 = \sqrt{MW}$. M is the mass of a nucleon, being nearly identical for neutron and proton, and W is the binding energy equal to $-E$, where E is negative for a bound state. Since ψ must be finite at $r = 0$, $A_1 = -A_2$, and Eq. (6.25) may be written

$$r < a: \quad u = A \sin(p_2 r/\hbar)$$
$$r > a: \quad u = B e^{-q_1 r/\hbar} \tag{6.25a}$$

where $u(r) = r\psi(r)$. Equation (6.25) may also be obtained from the Schroedinger wave equation in spherical coordinates.

In Chapter 8 it will be shown how a spherically symmetric problem may be rigorously reduced to a one-dimensional problem by making the substitution $u(r) = r\psi(r)$. Borrowing this result, we obtain the following wave equations:

$$r < a: \quad \frac{d^2u}{dr^2} + \frac{M}{\hbar^2}(V_0 - W)u = 0$$

$$r > a: \quad \frac{d^2u}{dr^2} - \frac{M}{\hbar^2}Wu = 0$$

(6.26)

The solutions are as given in Eqs. (6.25a) for boundary conditions requiring ψ to be finite at the origin and at infinity.

In order for u and its derivative to be continuous at $r = a$, the derivatives with respect to r of the logarithm of u on both sides of the boundary must be equal. Therefore,

$$p_2 \cot \frac{p_2 a}{\hbar} = -q_1$$

(6.27a)

that is,

$$-\sqrt{W/(V_0 - W)} = \cot(\sqrt{M(V_0 - W)}a/\hbar)$$

(6.27b)

If $W \ll V_0$, as in the lowest energy state, then

$$\sqrt{MV_0}\,a/\hbar \sim (\pi/2) - \sqrt{W/V_0}$$

Inserting typical theoretical values into the preceding equation, $a = 2.15 \times 10^{-13}$ cm and $V_0 = 30$ Mev, we find that $-W = E = -2.3$ Mev. Actually the binding energy of the deuteron has been known for some time to be 2.22 Mev. Application of Eq. (6.27) to obtain the best V_0 and a compatible with $W = 2.22$ Mev gave some of the best early estimates of the size and range of the potential of the neutron–proton interaction.

Problem 6.2. Determine the reflectivity and transmission for a particle with $E > 0$ incident from the left on a square well as illustrated in Figure 6.6.

Problem 6.3.   Using Feynman probability amplitude wave functions as in Problem 4.1, determine the transmission for the example of Problem 6.1 by first summing up all the probability amplitudes for the waves (1) transmitted at both surfaces ($x = 0$ and a), (2) transmitted at the front surface ($x = 0$), reflected from the back ($x = a$), and then from the front, and finally transmitted at $x = a$, (3) transmitted at $x = a$ after four internal reflections, (4) transmitted after six internal reflections, etc., for an infinite summation. Assume that the transmission and reflection coefficients for each encounter with a surface are given correctly by Eqs. (6.5).

Problem 6.4.   Find the energies available to an electron in a cube of metal $100\,\mu$ on a side if the work function for the metal is 5 ev. Idealize the problem by working in only one dimension.

Problem 6.5.   For $V_0 \gg W$, Eqs. (6.24) give the approximate result that

$$W = V_0 - \frac{n^2\pi^2\hbar^2}{2ma^2}$$

where n is an integer. If the force exerted by a particle on the sides of a one-dimensional square well is given by $F = \partial W/\partial a$, find the force that the particle exerts on a side of the well.

Problem 6.6.   Assuming the values for V_0 and a given in the text, show whether a second bound state for the deuteron, i.e., the first excited state, is possible. How can an unbound transient state arise if $E > 0$? (See Problem 6.2.)

6.3   THE BAND THEORY OF METALS

One of the best and simplest applications of a square well potential in one dimension is the application to the problem of determining the possible energy levels available to electrons in solids. The simple one-dimensional Cartesian coordinate system is sufficient for the analysis, and the long wavelength of the electrons compared to the distance between atoms in a crystal causes the electronic behavior to be independent in many respects of the exact shape of the potential. Hence, we should expect a simple square well potential to yield an excellent qualitative picture of the general behavior of electrons in solids.

In Problem 6.4 the electronic wave functions in a metal were calculated by assuming that a metal could be represented as a simple square well potential having the dimensions of the piece of metal. Motion in any one of the x, y, or z directions is assumed independent of motion in the other two directions. Therefore, the problem could be idealized by studying each motion separately and thus the problem was reduced to motion in one dimension. The solution, similar to that for the radiation problem of Chapter 1, consists of plane waves whose wave numbers ($k = 1/\lambda = p/\hbar$) must be proportional to an integral multiple of the reciprocal of the length of the sides of the piece of metal. Thus, the space-dependent parts of the possible electron wave functions are $e^{\pm ikx}$, where k is an integral multiple of π/d, d being the length of a side of the piece of metal. If the distance between metal atoms is a and there are N atoms along a side of length d (i.e., $d = Na$) then k must be equal to or less than $2\pi/a$ in order for the electrons to be represented by unique wave functions.

Figure 6.9 illustrates the problems one would encounter if one attempted to describe the positions of the N electrons having N degrees of freedom in their motion about the N atoms of the crystal by more than the N waves with wave numbers $k \leqslant 2\pi/a$. The very concept of wave motion implies a continuous rather than a discrete medium, and therefore the wavelengths must be equal to or greater than, not less than, the discreteness of the medium, i.e., the atomic separations, in order for a wave description to be valid. Therefore, $k = n\pi/Na$ where n is an integer less than $2N$. For a three-dimensional piece of metal the frequencies or energies are determined by the condition that $H\psi = -(\hbar^2/2m)\, \nabla^2\psi = E\psi$. Therefore,

$$E = \frac{\hbar^2\pi^2}{2m}\left[\left(\frac{n_1}{d_1}\right)^2 + \left(\frac{n_2}{d_2}\right)^2 + \left(\frac{n_3}{d_3}\right)^2\right] \tag{6.28}$$

where the subscripts refer to the three sides of the metal. (Note that since the Schroedinger wave equation is first order in time, E appears to the first power, rather than the frequency to the second power as for radiation in Chapter 1.) Thus, in close analogy to radiation in a box, the number of electronic wave functions having an energy between E and $E + dE$ is given by

$$N(E)\, dE = \frac{V}{4\pi^2}\left(\frac{2m}{\hbar^2}\right)^{3/2} \sqrt{E}\, dE \qquad (6.29)$$

where V is the volume of the piece of metal.

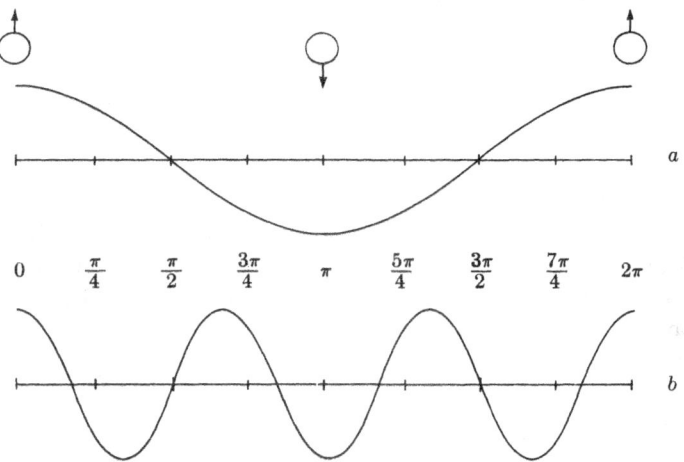

Figure 6.9   (a) *Illustration of the shortest wavelength that may be used to describe the motion of single valence electrons indicated above* (a). *If shorter wavelengths were permitted, the motion could be described by the wave* (b) *and an infinite number of higher frequency waves.*

The Pauli exclusion principle requires that only one electron at a time can occupy a single state described by a given wave function and spin polarization. (Since electrons may have spin with vector up or down, there are two directions of polarization just as for light, and when we include spin the total number of states is double the number given in Eq. 6.29.) The total number of electrons in the lowest total wave function is obtained by integrating Eq. (6.29) up to a maximum energy E_{max}, with the result

$$N = 2\frac{V}{4\pi^2}\left(\frac{2m}{\hbar^2}\right)^{3/2}(\tfrac{2}{3}E_{max}^{3/2})$$ (6.30)

or

$$E_{max} = \frac{3\pi^2\hbar^2}{2m}\left(\frac{N}{V}\right)^{2/3} = 36.1\left(\frac{N}{V}\right)^{2/3}$$

in units of electron volts if V is in cubic Ångstroms. This maximum energy which electrons in the ground state of a metal may have is called the Fermi energy. An electron having this energy is in the Fermi level. Hence even in a metal at zero degrees absolute temperature, with all electrons in the lowest possible states, some electrons will have kinetic energies of several electron volts.

Solids are, in actuality, more complicated than they have been represented in the foregoing. They consist of a lattice of atomic nuclei; hence, in place of a square well of constant depth, the electron potential well is a periodic function of distance with a period given by the atomic separation in the lattice. It is still possible, however, to simplify the problem by assuming the variables separable (which in actuality they are not) and then considering the wave motion in just one dimension. The Kronig–Penney potential energy model in one dimension for electrons in a crystal is illustrated in Figure 6.10. The crystal atoms form a periodic barrier of height V_0 and widths b spaced a distance a apart. Due to the periodicity of the potential the wave function must have a coefficient periodic in x. The x-dependent part of the wave function in Cartesian coordinates is made up of plane waves e^{ikx} (where k, an integral multiple of π/a, is less than twice the total number of atoms N along a side of the crystal) times a periodic function of period a. Let

$$\psi = u_k(x)\,e^{ikx}$$ (6.31)

where $u_k(x)$ is a general periodic function called a Bloch wave function for periodic potentials of any shape in a crystal.

Substitution of the potential into Eq. (4.8) yields time-independent standing wave solutions which are:

For $(n - 1)a \leqslant x \leqslant na - b$, where n is an integer $0 < n < N$:

$$\psi_k = A\, e^{i\alpha x} + B\, e^{-i\alpha x}$$
$$u_k = A\, e^{i(\alpha - k)x} + B\, e^{-i(\alpha + k)x} \qquad (6.32)$$

where

$$\alpha = p_1/\hbar = \sqrt{2mE/\hbar^2}$$

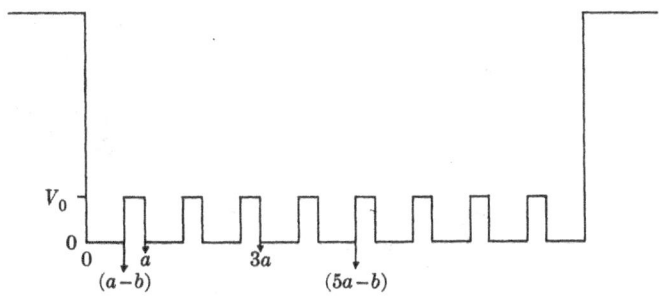

V_0

0

0 a $3a$

$(a-b)$ $(5a-b)$

Figure 6.10 *Illustration of the Kronig–Penney step-function barrier potential function representing the atoms in a crystal lattice.*

For $na - b \leqslant x \leqslant na$ the electron potential energy is increased by V_0 over the electron potential energy for $(n - 1)a \leqslant x \leqslant na - b$, and the electron wave number changes from $\sqrt{2mE/\hbar^2}$ to $i\sqrt{2m(V_0 - E)/\hbar^2}$. For $na - b \leqslant x \leqslant na$ and $V_0 > E$:

$$\psi_k = C\, e^{\beta x} + D\, e^{-\beta x}$$
$$u_k = C\, e^{(\beta - ik)x} + D\, e^{-(\beta + ik)x} \qquad (6.33)$$

where

$$\beta = p_2/\hbar = \sqrt{2m(V_0 - E)/\hbar^2}$$

Equations (6.32, 6.33) are customarily found as solutions to the complete wave equation in which the Bloch wave function Eq. (6.31) has been substituted, which is

$$\frac{d^2 u_k}{dx^2} + 2ik\,\frac{du_k}{dx} + \left[\frac{2m}{\hbar^2}(E - V) - k^2\right]u_k = 0 \qquad (6.34)$$

The conditions requiring that u and its first derivative be continuous in general at $x = (n - 1)a$ and $x = na - b$ and in particular at $x = 0$ and $x = a$, *and the particular periodicity condition* $u(a) = u(0)$, $\psi(a \pm b) = \psi(-b)$, yield the four equations

$$A + B = C + D$$

$$i(\alpha - k)A - i(\alpha + k)B = (\beta - ik)C - (\beta + ik)D$$

$$A\, e^{i(\alpha - k)(a-b)} + B\, e^{-i(\alpha + k)(a-b)} = C\, e^{-(\beta - ik)b} + D\, e^{(\beta + ik)b}$$

$$i(\alpha - k)A\, e^{i(\alpha - k)(a-b)} - i(\alpha + k)B\, e^{i(\alpha + k)(a-b)}$$
$$= (\beta - ik)C\, e^{-(\beta - ik)b} - (\beta + ik)D\, e^{(\beta + ik)b}$$

Either A, B, C, and D are all simultaneously zero or the determinant of the coefficients of these four simultaneous algebraic equations must vanish. Setting the determinant equal to zero, going through a great deal of algebraic manipulation and inserting $\sin \alpha(a - b)$ for $(1/2i)(e^{i\alpha(a-b)} - e^{-i\alpha(a-b)})$, $\cos \alpha(a - b)$ for $\frac{1}{2}(e^{i\alpha(a-b)} + e^{-i\alpha(a-b)})$, $\sinh \beta b$ for $\frac{1}{2}(e^{\beta b} - e^{-\beta b})$, and $\cosh \beta b$ for $\frac{1}{2}(e^{b} + e^{-})$, we find

$$\frac{\beta^2 - \alpha^2}{2\alpha\beta} \sinh \beta b \sin \alpha(a - b) + \cosh \beta b \cos \alpha(a - b) = \cos ka \qquad (6.35)$$

For simplicity, we may replace our potential barrier by a delta function so constructed that

$$\lim_{\substack{b \to 0 \\ \beta \to \infty}} \beta^2 ab = 2P$$

where P is some constant, and use $\sin \alpha(a - b) = \sin \alpha a$, $\cos \alpha(a - b) = \cos \alpha a$, $\cosh \beta b = 1$, and $((\beta^2 - \alpha^2)/2\alpha\beta) \sinh \beta b = (\beta/2\alpha)\beta b$, so that Eq. (6.35) becomes

$$(\beta/2\alpha)(\beta b) \sin \alpha a + \cos \alpha a = \cos ka$$

whence

$$[(P \sin \alpha a)/\alpha a] + \cos \alpha a = \cos ka \qquad (6.36)$$

The left-hand side of Eq. (6.36) must be less than or equal to one in order for k to be real and not give rise to rapidly decaying exponential wave functions. A plot of the left-hand side of (6.36) for $P = 3\pi/2$ as a function of αa is shown in Figure 6.11. The shaded regions are for those values of αa for which solutions do not exist. Figure 6.12 illustrates the allowed values of $\alpha^2 a^2 = 2mEa^2/\hbar^2$ in Eq. (6.36) as a function of ka. Note the energy gaps occurring at integral multiples of $\pi = ka$, separating the bands of allowed energy levels. Since $E \sim k^2$, the allowed energies have a roughly parabolic dependence on ka. Portions of the three lowest eigenfunctions

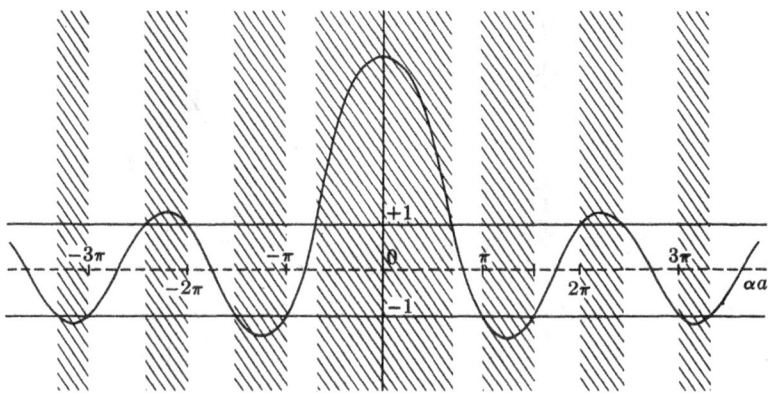

Figure 6.11 *Graph of $[(P \sin \alpha a)/\alpha a] + \cos \alpha a$ as a function of αa. The dashed lines $+1$ and -1 are the maximum values the function can have in possible solutions, hence the shaded areas are unallowed values for αa.*

which occur in a Kronig–Penney potential are shown approximately in Figure 6.8. The exact wave functions in this potential are slightly shifted from those in the figure since the wave function in Eq. (6.33) includes an exponentially increasing term.

The electrons in a metal about half fill the conduction band. An applied electric field causes the electrons to accelerate in the direction of the field, in between collisions with the lattice vibrations. In this way a net flow of electricity takes place in the metal. In other substances, however, the energy bands are either completely empty or completely filled at ordinary temperatures. An electric field applied to nonmetals cannot accelerate the electrons, for acceleration or falling through the applied electric field would raise the energy of the electron, which is not permitted unless the energy is raised discontinuously so that the electron jumps to an unfilled energy band. (Note that the Pauli principle forbids the electron from entering a state which is already occupied.)

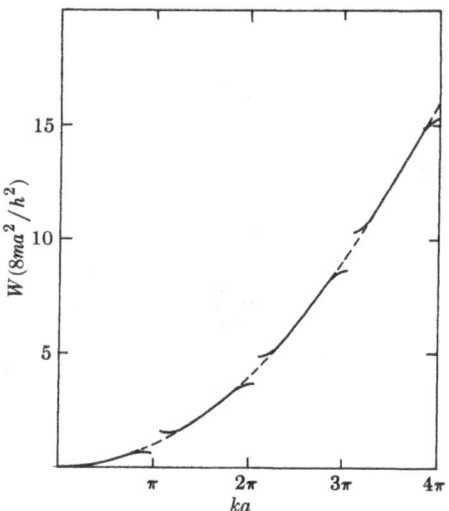

Figure 6.12 *Graph of energy as a function of wave number for the Kronig–Penney potential.*

Problem 6.7 Derive Eq. (6.29) by the methods given in Chapter 1.

BIBLIOGRAPHY

Rojansky, Vladimir, *Introductory Quantum Mechanics*. Englewood Cliffs, N.J.: Prentice-Hall, Inc., 1946.

Bohm, David, *Quantum Theory*. Englewood Cliffs, N.J.: Prentice-Hall, Inc., 1951.

Mott, N. F., and H. Jones, *The Theory of the Properties of Metals and Alloys*. New York: Dover Publications, Inc., 1958.

Kittel, Charles, *Introduction to Solid State Physics*. New York: John Wiley & Sons, Inc., 1953.

Bethe, Hans A., and Phillip Morrison, *Elementary Nuclear Theory*. New York: John Wiley & Sons, Inc., 1956.

THE WKB APPROXIMATION AND THE

BOHR-SOMMERFELD QUANTUM CONDITIONS

IN THE LAST CHAPTER, to emphasize the nonclassical aspects of particle behavior revealed by quantum mechanics, we treated the extreme case of potentials which change discontinuously along a particle path, i.e., square potentials. Since classical physics is known to be an excellent approximation in many instances, it is instructive to see how classical mechanics may arise as an approximation to our more general theory. The wave behavior of particles has been shown to arise from the Heisenberg uncertainty principle, $\Delta p \Delta x \geqslant \hbar$, which would agree with classical physics if $\hbar = 0$. This suggests that we should look for a series expansion in powers of $\hbar$ in our probability amplitude solutions to the wave equation. We should further expect this method of approximating wave equation solutions to be very useful when the potential energy of the particle changes slowly within a particle wavelength, in contrast to the discontinuous changes treated in the previous chapter.

Up to this point we have examined solutions of the wave equation in regions where V (and therefore p) was a constant. In these circumstances the particle wavelength could be and was unique. This was possible only because no questions were asked about the particle position. In regions where the potential energy is a function of position, however, the wave function clearly could not be written as a simple plane wave with unique momentum. The wave function $\exp\{(i/\hbar)\int p\,dx\}$ corresponds to a sort of frequency-modulated wave; and in a slowly varying potential the sum of this wave function over all possible paths might be approximated by the constant potential wave form times a more slowly varying fudge factor. We will now show that just such an approximate wave function will result from an attempt to expand our wave function in increasing powers of $\hbar$.

7.1 THE WENTZEL-KRAMERS-BRILLOUIN METHOD

Letting $p = \sqrt{2m(E - V)}$, we can rewrite the one-dimensional time-independent wave equation as

$$\frac{d^2\psi}{dx^2} + \frac{p^2}{\hbar^2}\psi = 0 \tag{7.1}$$

Since p is now a function of position, the solution is no longer of the form $e^{(i/\hbar)px}$, but this form will be shown to be a reasonable first approximation when certain conditions are met. We first choose a new dependent variable, u:

$$u = \frac{\hbar}{i}\frac{1}{\psi}\frac{d\psi}{dx}, \qquad \psi = \exp\left(\frac{i}{\hbar}\int^x u\,dx\right) \tag{7.2}$$

so that Eq. (7.1) becomes

$$\frac{\hbar}{i}\frac{du}{dx} = p^2 - u^2 \tag{7.3}$$

We seek solutions whose first approximation give the classical results, by expanding u in powers of $\hbar/i$,

$$u = u_0 + \frac{\hbar}{i}u_1 + \left(\frac{\hbar}{i}\right)^2 u_2 + \ldots \tag{7.4}$$

Substituting (7.4) into (7.3) and keeping terms of order no higher than $(\hbar/i)$ we obtain

$$\frac{\hbar}{i}\frac{du_0}{dx} = p^2 - u_0{}^2 - 2\frac{\hbar}{i}u_1u_0 \qquad (7.5)$$

The zero-order approximation amounts to neglecting $(\hbar\,du/i\,dx)$ in Eq. (7.3). Equating coefficients of equal powers of $\hbar/i$ we obtain

$$u_0 = \pm\,p, \qquad u_1 = -\frac{1}{2}\frac{1}{u_0}\frac{du_0}{dx}$$

leading to the two alternative solutions

$$u_+ = p - \frac{\hbar}{i}\frac{1}{2p}\frac{dp}{dx} = p - \frac{\hbar}{i}\frac{d}{dx}\log_e\sqrt{p}$$

$$u_- = -p - \frac{\hbar}{i}\frac{1}{2p}\frac{dp}{dx} = -p - \frac{\hbar}{i}\frac{d}{dx}\log_e\sqrt{p} \qquad (7.6)$$

Substituting Eq. (7.6) into Eq. (7.2) we find the approximate wave function

$$\psi = \frac{A}{\sqrt{p}}\exp\left(\frac{i}{\hbar}\int_{x_1}^{x} p\,dx\right) + \frac{B}{\sqrt{p}}\exp\left(-\frac{i}{\hbar}\int_{x_1}^{x} p\,dx\right) \qquad (7.7)$$

In the limit of short wavelength, i.e., large momenta, Eq. (7.7) resembles a plane wave with phase given by the action, $\int p\dot{q}\,dt = \int p\,dq$, in units of $\hbar$. As $\hbar/p$ approaches zero, the wave function given in Eqs. (7.2, 7.7) becomes a rapidly oscillating function of x so that noncancelling contributions to the total wave function from neighboring paths occur only in that region of x where the action varies least rapidly with change in path. This region, where the action is an extremum, is also the region for which Hamilton's principal function, $\int L\,dt$ (where L is the Lagrangian, Chap. 4), is an extremum. The fact that a light ray follows that path for which the action is also either a minimum or a maximum is known as Fermat's principle in optics. It is the classical path taken by a particle or a light ray. The Wentzel–Kramers–Brillouin or WKB method has given us this solution plus additional terms beyond the classical expression, of which we have taken only the first. This additional term in the exponent results in a factor $p^{-1/2}$ in the wave function. With this change, the probability of finding a particle per unit volume, $\psi^*\psi$, becomes $1/p$ times the constant-potential probability. The time a particle spends in any one spot is inversely proportional to its velocity, and the first-order correction to the plane wave case has given us just this result, which is also entirely in accord with classical mechanics. For example, the bob in a clock pendulum is most likely to be found at any given instant at the end of its swing where its velocity is zero.

The series expansion expressed in Eq. (7.4) is an asymptotic series which diverges after the first few terms. The first few terms give a good approximation, however, as long as

$$\frac{\hbar \, dp}{p^2 \, dx} \ll 1$$

This condition, which is seen to follow from requiring that the second term in Eq. (7.6) be much smaller than the first term, is the usual condition for validity in ray optics or classical physics; namely that the wave number of the wave function ($\hbar/p$ for particles, ω/nc for light where n is the index of refraction) experience little change over one wavelength. Put more precisely, the condition is that the fractional change in a wavelength within a distance equal to the wavelength be much less than unity. As the wavelength grows larger, that is, as p approaches zero, u_1, u_2, and higher terms in the expansion all diverge, as is indicated in Eq. (7.6), and the approximation is no longer valid; in this case, in general, one does better to go to the other extreme and seek a long-wavelength or physical-optics type of solution to the Schroedinger equation. That is, it is preferable to solve a square well or barrier problem in first approximation.

We now apply our method to the potential illustrated in Figure 7.1, for which $p = \sqrt{2m(E - V)}$ is imaginary in regions I and III and real in region II. Since ψ must be finite everywhere and p_I and p_{III} are imaginary, u_+ is an unacceptable solution for $x < x_1$ and u_- is unacceptable for $x > x_2$. Only exponentially decreasing solutions are possible in regions I and III for which, respectively, $A_I = 0$ and $B_{III} = 0$. In region II both positive and negative exponential solutions are acceptable and may be combined and rewritten as

$$\psi_{II} = \frac{2C_{II}}{\sqrt{p}} \cos\left(\frac{1}{\hbar} \int_{x_1}^{x} p \, dx + \delta\right) \tag{7.8}$$

with

$$A_{\mathrm{II}} = C_{\mathrm{II}}\, e^{i\delta} \quad \text{and} \quad B_{\mathrm{II}} = C_{\mathrm{II}}\, e^{-i\delta}$$

whereas

$$\psi_{\mathrm{I}} = (B_{\mathrm{I}}/\sqrt{i\overline{p}})\exp\left(\frac{i}{\hbar}\int_{x_1}^{x} ip\,dx\right), \qquad \psi_{\mathrm{III}} = (A_{\mathrm{I}}/\sqrt{i\overline{p}})\exp\left(\frac{i}{\hbar}\int_{x_2}^{x} ip\,dx\right)$$

Conditions will be imposed on these solutions in the next section.

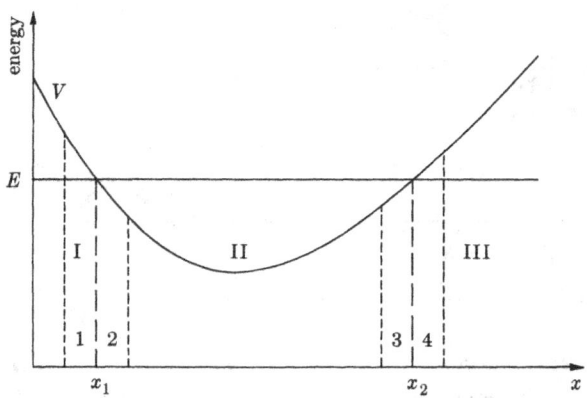

Figure 7.1 *Energy graph for a particle with energy E bound in a potential well for which $E < V$ in $x < x_1$ and $x > x_2$.*

7.2 CONNECTION FORMULAS

The solutions in the three regions must now be matched at x_1 and x_2 by the usual conditions of continuity. The difficulty is that at the points x_1 and x_2, $p = 0$, and Eq. (7.7) becomes an unsatisfactory approximation to the wave equation. At these points V may be approximated as a linear function of x by the first term in a Taylor's series expansion of V about the points $x = x_1$ and $x = x_2$. Hence the wave equation near x_1 may be written

$$\frac{d^2\psi}{dx^2} + n(x - x_1)\psi = 0 \qquad (7.9)$$

where n is $2m/\hbar^2$ times the negative slope of the potential at x_1, that is, $(2m/\hbar^2)(-\partial V/\partial x)$, where $-\partial V/\partial x$ is evaluated at $x = x_1$. The solution of Eq. (7.9) is exact, and to the right of x_1 it is

$$\psi = D\sqrt{x - x_1}J_{\frac{1}{3}}(\tfrac{2}{3}\sqrt{n}[x - x_1]^{3/2}) + E\sqrt{x - x_1}J_{-\frac{1}{3}}(\tfrac{2}{3}\sqrt{n}[x - x_1]^{3/2}) \quad (7.10)$$

where $J_{\frac{1}{3}}$ is the Bessel function of order $\frac{1}{3}$. A graph of $J_{\frac{1}{3}}(x)$ and $J_{-\frac{1}{3}}(x)$ is shown in Figure 7.2.*

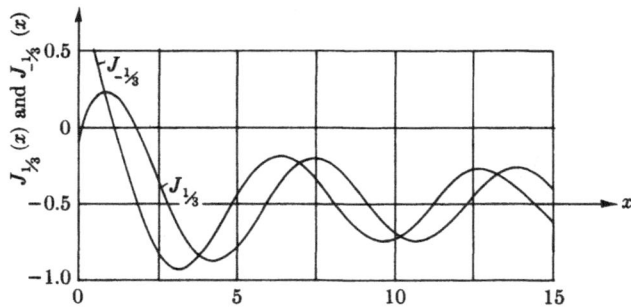

Figure 7.2 *Graph of* $J_{\frac{1}{3}}(x)$ *and* $J_{-\frac{1}{3}}(x)$. *Note that for large x both functions become sinusoidal with slowly decreasing amplitude and with period equal to* 2π. (*From Jahnke, Emde, and Lösch,* Tables of Higher Functions, *Stuttgart : B. G. Teubner Verlagsgesellschaft,* 1960).

As x increases without limit, Eq. (7.10) asymptotically approaches a cosine function:

$$\psi \simeq \sqrt{\frac{3}{\pi n^{1/2}}} \frac{D}{(x - x_1)^{1/4}} \cos\left[\tfrac{2}{3}\sqrt{n}(x - x_1)^{3/2} - \frac{5\pi}{12}\right]$$

$$+ \sqrt{\frac{3}{\pi n^{1/2}}} \frac{E}{(x - x_1)^{1/4}} \cos\left[\tfrac{2}{3}\sqrt{n}(x - x_1)^{3/2} - \frac{\pi}{12}\right]$$

$$(7.11a)$$

* For a description of Bessel functions see a calculus text such as *Methods of Advanced Calculus*, P. Franklin, McGraw-Hill, N.Y. (1944) or *Applied Mathematics for Engineers and Physicists*, L. A. Pipes, McGraw-Hill, N.Y. (1946).

To the left of x_1 the same solution (7.10) results except that the argument of the Bessel function is imaginary; their asymptotic expansion for large absolute value of $(x - x_1)$ includes positive and negative exponentials:

$$\psi \simeq \sqrt{\frac{3}{4\pi n^{1/2}}} \frac{D}{(x - x_1)^{1/4}}[e^{(2/3)\sqrt{n}(x_1-x)^{3/2}} + e^{-i\pi/6}\, e^{-(2/3)\sqrt{n}(x_1-x)^{3/2}}]$$

$$- \sqrt{\frac{3}{4\pi n^{1/2}}} \frac{E}{(x-x_1)^{1/4}}[e^{(2/3)\sqrt{n}(x_1-x)^{3/2}} + e^{-5i\pi/6}\, e^{-(2/3)\sqrt{n}(x_1-x)^{3/2}}]$$

$$(7.11b)$$

These asymptotic expressions (for the linear potential approximation at the turning points) must fit the solutions we have already found for regions I, II, and III. In order for ψ to be zero at $-\infty$, the WKB solution in region I must be

$$\psi_{\mathrm{I}} = (B_{\mathrm{I}}/\sqrt{-ip})\exp\left(-\frac{i}{\hbar}\int_{x_1}^{x} p\,dx\right) \qquad (7.12)$$

where

$$p = \sqrt{2m(E - V)}$$

This solution matches Eq. (7.11b) if

$$D = E = \frac{\sqrt{4\pi/3\hbar}}{2\sin(\pi/3)} B_{\mathrm{I}}$$

since

$$e^{-i\pi/6} - e^{-5i\pi/6} = e^{-i\pi/2}\, 2i\sin(\pi/3)$$

For this value of D and E Eq. (7.11a) becomes

$$\psi = \frac{2}{[n\hbar^2(x - x_1)]^{1/4}}\frac{\cos(\pi/6)}{\sin(\pi/3)} B_{\mathrm{I}}\cos\left[\tfrac{2}{3}\sqrt{n}(x - x_1)^{3/2} - \frac{\pi}{4}\right] \qquad (7.13)$$

since

$$\cos\left(a - \frac{5\pi}{12}\right) + \cos\left(a - \frac{\pi}{12}\right) = 2\cos\left(a - \frac{\pi}{4}\right)\cos\frac{\pi}{6}$$

Equation (7.8) matches Eq. (7.13) if $C_{\mathrm{II}} = B_{\mathrm{I}}$, giving

$$\psi_{\mathrm{II}} = \frac{2B_{\mathrm{I}}}{\sqrt{p}}\cos\left(\frac{1}{\hbar}\int_{x_1}^{x} p\,dx - \frac{\pi}{4}\right) \qquad (7.14)$$

Similarly, if

$$\psi_{\mathrm{III}} = \frac{A_{\mathrm{III}}}{\sqrt{-ip}}\exp\left(\frac{i}{\hbar}\int_{x_2}^{x}p\,dx\right) \tag{7.15}$$

then

$$\psi_{\mathrm{II}} = \frac{2A_{\mathrm{III}}}{\sqrt{p}}\cos\left(\frac{1}{\hbar}\int_{x}^{x_2}p\,dx - \frac{\pi}{4}\right) \tag{7.16}$$

If Eqs. (7.14, 7.16) are to be the same solution, then $B_{\mathrm{I}} = A_{\mathrm{III}}$ and the absolute values of the arguments of the cosine must differ only by an integral multiple of π. That is,

$$\left|\frac{1}{\hbar}\int_{x_1}^{x}p\,dx - \frac{\pi}{4}\right| = \left|\frac{1}{\hbar}\int_{x}^{x_2}p\,dx - \frac{\pi}{4}\right| + n'\pi$$

or

$$\frac{1}{\hbar}\int_{x_1}^{x}p\,dx - \frac{\pi}{4} = \pm\left(\frac{1}{\hbar}\int_{x}^{x_2}p\,dx - \frac{\pi}{4}\right) + n\pi \tag{7.17}$$

Since this condition must be satisfied for all x the minus sign must be taken, and hence the permissible values of p and therefore E are given by

$$\frac{1}{\hbar}\int_{x_1}^{x_2}p\,dx = (n + \tfrac{1}{2})\pi \tag{7.18}$$

That is, the energy is quantized in accordance with Eq. (7.18); only certain discrete values are possible. As stated previously, *only discrete values of the energy are ever possible for particles confined to a finite volume.*

The important consequence of the WKB method of solving the Schroedinger equation for a bound state problem is that simple quantum conditions are imposed on the possible energy states of the system. A simple formula, Eq. (7.18), has been found for obtaining them. Note that this result is valid only when the potential does not change much within a wavelength and in cases where we may obtain solutions far from the points where $p = 0$. This means, as we should expect, that the approximation is valid only for slowly varying potentials and in the limit of large quantum numbers where the classical result $\Delta E/E = 0$, i.e., no uncertainty in the energy, is a reasonably good first approximation.

7.3 THE BOHR–SOMMERFELD QUANTUM CONDITIONS

Equation (7.18) may be rewritten

$$\oint p \, dx = (n + \tfrac{1}{2})h. \tag{7.19}$$

since twice the path integral from x_1 to x_2 is equivalent to a complete cycle of the particle motion in its oscillation between the classical limits of its motion. This approximate quantization rule (for p and thus E) is identical, but for the extra term $h/2$, with the formula Bohr and Sommerfeld postulated a decade before the formulation of wave mechanics to predict the energy levels possible for a bound particle. For low quantum numbers the extra term $h/2$ gives a better approximation.

As originally stated the Bohr–Sommerfeld rule was intended to apply to any periodic motion, including rotation, whereas the above derivation applies only to vibrational motion. The corresponding rule for rotation is that the integral of the angular momentum over a complete cycle must yield an integer, not half an integer, since V and hence p_θ must have an angular period of 2π for rotational motion, where p_θ is everywhere real. That, is, for rotational motion where θ is the angular displacement and p_θ the angular momentum.

$$\oint p_\theta \, d\theta = nh \tag{7.19a}$$

As an example of the utility of Eq. (7.19) we may compare the level density, or the number of energy levels per unit energy, for the most energetic nucleons in actual heavy nuclei having a diffuse wall, with the estimated level density in a square well. If the modification resulting from the three-dimensional aspect of the problem is neglected, the energy levels in a square well are given by the phase condition at the wall, Eq. (6.23):

$$\varphi = \left(\sqrt{\frac{m}{2}(V_0 - W_n)} \frac{a}{\hbar} \right) + \left(\frac{n\pi}{2} \right) \sim \sqrt{\frac{W_n}{V_0}} \tag{7.20}$$

where $W_n = -E_n$ and $\varphi = \tan^{-1} \sqrt{W_n/(V_0 - W_n)}$. The right-hand side is an approximation valid when W_n is much less than V_0. The approximate solution is

$$W_n = V_0 - (h^2 n^2 / 8ma^2) \qquad (7.21)$$

The level density in the square well is given by the reciprocal of the level spacing, ΔE, between levels of neighboring n:

$$\frac{1}{\Delta E} = \frac{1}{E_{n+1} - E_n} = \frac{8ma^2/h^2}{2n+1} \sim \frac{\sqrt{2m}\, a/h}{\sqrt{V_0 - W_n}} \qquad (7.22)$$

since

$$n = \sqrt{8m}(a/h)\sqrt{V_0 - W_n}$$

(Note that for large n, ΔE is given by $\Delta E \sim \partial E/\partial n$.)

For the high quantum numbers (resulting in $W_n \ll V_0$) characteristic of the most energetic nucleons in heavy nuclei, the number of energy levels per unit energy in a square well would be $\sqrt{2ma^2/V_0 h^2}$, a constant. In a realistic diffuse well, however, as has been deduced from nucleon scattering experiments, the nuclear potential is better described by an Eckart–Bethe or Woods–Saxon potential,

$$V = -V_0(1 + e^{(r-a)/b})^{-1}$$

Substitution of this potential into Eq. (7.19) leads to the result

$$h(n + \tfrac{1}{2}) = 2 \int_0^{r_2} \sqrt{2m[E_n + V_0(1 + e^{(r-a)/b})^{-1}]}\, dr$$

where r_2 is defined by

$$1 + e^{(r_2-a)/b} = -V_0/E_n$$

Let

$$x = [1 + e^{(-r-a)/b}]^{-1}$$

from which

$$r = a + b \log_e \frac{1 - x}{x}$$

then

$$h(n + \tfrac{1}{2}) = 2\sqrt{2m}\, b \int_{(1+e^{-a/b})^{-1}}^{-E_n/V_0} \sqrt{E_n + V_0 x}\left(\frac{1}{x} + \frac{1}{1-x}\right) dx$$

$$= -4\sqrt{2m}\, b\Bigg\{ -\sqrt{E_n + V_0/(1 + e^{-a/b})}$$

$$+ \sqrt{E_n + V_0[1 - (1 + e^{a/b})^{-1}]}$$

$$+ \sqrt{-E_n}\,\tan^{-1}\sqrt{-1 - V_0/(1 + e^{-a/b})E_n}$$

$$+ \tfrac{1}{2}\sqrt{E_n + V_0}\log_e \left| \frac{\sqrt{E_n + V_0[1 - (1 + e^{a/b})^{-1}]} - \sqrt{E_n + V_0}}{\sqrt{E_n + V_0[1 - (1 - e^{a/b})^{-1}]} + \sqrt{E_n + V_0}} \right| \Bigg\}$$

$$\tag{7.23}$$

which can be approximated as

$$h(n + \tfrac{1}{2}) \underset{|E_n| \ll V_0}{\approx} -4\sqrt{2m}\, b\left\{ \frac{\pi}{2}\sqrt{-E_n} - \tfrac{1}{2}\sqrt{E_n + V_0}\left[\frac{a}{b} + (\log_e 4)\frac{E_n}{V_0}\right]\right\}$$

from which the energy level density in actual heavy nuclei for high quantum numbers may be readily deduced in the same way as that for a square well:

$$\frac{1}{\Delta E} = \frac{\partial n}{\partial E_n} = \frac{4\sqrt{2m}\, b}{h}\left\{ \frac{\pi}{4}\frac{1}{\sqrt{-E_n}} + \frac{1}{4}\left[\frac{a}{b} + (\log_e 4)\right]\frac{1}{\sqrt{V_0}}\right\}$$

$$= \frac{C_1}{\sqrt{-E_n}} + \frac{C_2}{\sqrt{V_0}} \tag{7.24}$$

which reduces to $\sqrt{2m/V_0}\, a/h$, the square well result, if $b = 0$. The constants C_1 and C_2 are approximately 1.5 and 3.1 (Mev)$^{-1/2}$, respectively, for the nuclear potential $V_0 \sim 42$ Mev and $|E_n| \ll V_0$ for the region of interest. The square well result ($\sim C_2/\sqrt{V_0}$) is independent of energy. Hence a square well, which is such a useful approximation in many nuclear physics applications, turns out sometimes to be a very bad approximation, this time for the energy level densities in highly excited heavy nuclei, to which it has frequently been applied. The extra term $C_1/\sqrt{-E_n}$ which distinguishes this calculation from that for a square well is significant for high quantum numbers where $|E_n| \ll V_0$. One should expect that in the limit of high quantum numbers and short wavelengths the square well approximation, so useful for long wavelengths, would break down. The kinetic energies of nucleons in these levels is of the order of 40 Mev, giving them a wavelength of about 0.7 fermi (one fermi $= 10^{-13}$ cm). While the core of the nucleus is indeed a region of constant potential, the distance in which the nuclear potential changes from this constant interior value to zero is about 0.8 fermi, and thus the potential change will not appear square to particles havnig a wavelength of around 0.7 fermi.

Problem 7.1 In the next chapter it will be shown that for central (angle-independent) potentials, the wave equation for the wave function for the radial coordinate may be written

$$\frac{d^2u}{dr^2} + \frac{2m}{\hbar^2}\left[E - V(r) - \frac{\hbar^2}{2m}\frac{l(l+1)}{r^2}\right]u = 0 \qquad (8.29)$$

where l is an integer and u is r times the radial wave function. ($\hbar^2 l(l+1)$ is the eigenvalue of the square of the angular momentum operator, and

$$\frac{\hbar^2}{2m}\frac{l(l+1)}{r^2}$$

represents the rotational energy of a particle about the origin.) Using the WKB method (that is, the quantization condition resulting therefrom), compute the number of energy levels per unit energy available to a nucleon for a particular value of $l = L$ in a nuclear square well of radius 10^{-12} cm and well depth 42 Mev. (Hint: the radial component of the momentum is given by $p_r^2 = 2m[E - V(r) - \hbar^2 l(l+1)/2mr^2]$.)

Problem 7.2 Using the WKB approximation, find the permissible levels for hydrogen for a particular value of $l = L$. The central potential is $V(r) = -e^2/r$. What is the total number of bound state levels as a result of the great extent of the Coulomb potential?

Problem 7.3 Using the WKB approximation, find the permissible levels for a harmonic oscillator in three dimensions, with $V(r) = +\frac{1}{2}Kr^2$, for a particular value of $l = L$.

Problem 7.4 By the WKB method find the energy levels in a one-dimensional well with $V(x) = C|x|$.

7.4 PENETRATION THROUGH A BARRIER

We now turn our attention to another historical application of the WKB approximation. Instead of the valley shown in Figure 7.1 we now have the hill of Figure 7.3. As in the derivation of Eq. (7.18), we first find that to the right of the barrier (which we label region III) $B_{III} = 0$, with ψ_{III} representing a wave traveling to the right. The wave function inside the barrier,

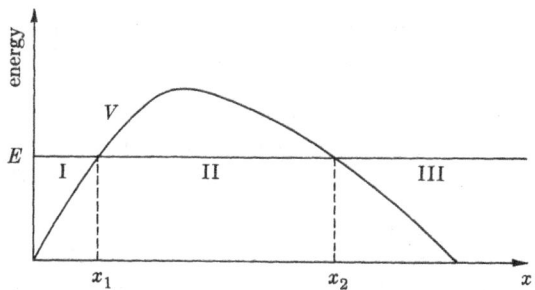

Figure 7.3 *Energy graph for a particle with energy E incident from the left on a barrier for which $E < V$ for $x_1 < x < x_2$.*

ψ_{II}, is made up of exponentials of real positive and negative argument which must be matched to the incident and reflected wave in region I (see Chapter 12 of Bohm, cited here in Chapter 6). Just as in Chapter 6, the transmission through the barrier is found from the ratio of the probability or intensity, $\psi^*\psi$, of the transmitted wave to the intensity of the incident wave times the ratio of the velocity in region III to the velocity in region I. Since the velocities are the same, the result is that

$$T = \exp - \left(\frac{2}{\hbar} \int_{x_1}^{x_2} p\,dx \right) \tag{7.25}$$

where

$$p = \sqrt{2m(V - E)}$$

Note that except for the refractive-index-like factor $(q_2/p_1 + p_1/q_2)^{-2}$ in front, which is negligible for the large values of the exponent that are required for the validity of the WKB approximation, Eq. (7.25) is similar to Eq. (6.12) used in calculating alpha particle decay. We should expect Eq. (7.25) to give a much better estimate of alpha particle half-lives than Eq. (6.12). We may approximate the barrier potential by a simple repulsive Coulomb potential beyond the attractive nuclear square well potential. That is, for an alpha particle of charge $2e$ and a residual nucleus of charge Ze,

$$V = 2Z\,e^2/r, \quad r > x_1 \tag{7.26}$$

where the radius of the alpha particle–nucleus attraction is given by

$$x_1 = (1.3 + 1.35A^{1/3}) \cdot 10^{-13}\,\text{cm}$$

where A is the atomic mass number of the residual nucleus.* The point at which the alpha particle energy exceeds the Coulomb potential is

$$x_2 = 2Ze^2/E$$

For these values of V, x_1, and x_2, Eq. (7.25) becomes

$$T = \exp\left\{ -\frac{2}{\hbar} \int_{x_1}^{x_2} \sqrt{2m[(2Ze^2/r) - E]}\,dr \right\}$$

$$= \exp\left\{ \frac{-8mZe^2}{\hbar\sqrt{2mE}} \tan^{-1}\left[\sqrt{\frac{x_2}{x_1} - 1} - \sqrt{\frac{x_1}{x_2}\left(1 - \frac{x_1}{x_2}\right)} \right] \right\}$$

$$\tag{7.27}$$

In the case of Po^{210}, $T \simeq e^{-64} \simeq 10^{-27}$. If the estimated frequency of interaction with the walls derived in Chapter 6, $10^{21}/\text{sec}$, is used, a half-life of $(1/T) \times 10^{-21} = 10^6$ sec results. This is to be compared with the measured half-life of 138 days or $\sim 10^7$ sec. Since the calculation is approximate (the estimated frequency of interaction with the wall is crude, and we have neglected the nuclear well shape and the reduction in transmission due to refractive-index-type factors), the results may be regarded as in rather good agreement with the observed half-life.

Equation (7.25) has another similar application, to the theory of the Schottky effect, i.e., electron emission from metals under an applied potential.

* J. O. Rasmusson, *Revs. Modern Phys.*, **30**, 424 (1958).

For this case $V = W - eEx$ for $x > 0$ where E is the applied electric field, x is the distance measured from the surface of the metal, and W is the work function for the metal. From (7.25), the rate of electron evaporation is thus proportional to

$$T = \exp\left[-\frac{2}{\hbar} \int_0^{x_2} \sqrt{2m(W - eEx)}\, dx \right] = \exp\left[-\tfrac{4}{3}\sqrt{2m}\,\frac{W^{3/2}}{\hbar eE} \right] \quad (7.28)$$

where x_2 is the point where $W = eEx$. Cold emission of electrons from metals exhibits just such a sensitive dependence on the work function of the metal and the applied electric field.

Problem 7.5 What are the half-lives of U^{238}, Np^{235}, and Po^{213}? Consult the *General Electric Chart of the Nuclides* for alpha particle energies, atomic mass numbers, and observed half-lives. Note that even if the absolute value of the result is only approximate, the ratio of one half-life to another fits the observed values rather well.

Problem 7.6 What are the barrier penetrabilities of Li, Fe, and Pb for 10 Mev protons, of Fe for 2 Mev protons, and of Li, Fe, Pb, for 10 Mev deuterons and alpha particles? (Compare the estimates of the penetrability with the curves of P. Morrison (see Bibliography of this chapter).)

Problem 7.7 If the work function in a metal is 3 ev and the applied electric fields are 10^3, 10^4, and 10^5 volts/m, what is the cold emission current for each case if 10^{28} conduction electrons arrive per cm^2-sec at the surface?

Problem 7.8 What is the penetrability of a particle through a barrier $V = V_0 - c|x|$?

Problem 7.9 What is the penetrability of an alpha particle with energy E through a nuclear square well and Coulomb barrier if the particle has an angular momentum of $10\hbar$ (i.e., $l = 10$)? (See Problem 7.1.)

Problem 7.10 Ordinarily the cold emission of metals is so affected by surface roughness that the simple WKB calculation is adequate. However, the problem can be solved exactly. On the side of the barrier where the potential changes rapidly, refractive-index-type factors result which may significantly improve the result. Obtain the exact expression for the transmission probability of an electron with wave number $k = p/\hbar$ incident from the left on a barrier changing abruptly at $x = -d$ from a constant value, $-V_0$, for $-d > x$ to $V = -Ax$ for $-d < x$ $(E = 0)$.

BIBLIOGRAPHY

Persico, E. (trans. by G. Temmer), *Fundamentals of Quantum Mechanics.* Englewood Cliffs, N.J.: Prentice-Hall, Inc., 1946.

Schiff, L., *Quantum Mechanics.* New York: McGraw-Hill Book Co., 1955.

Morrison, P., in *Experimental Nuclear Physics*, Vol. 2, ed. E. Segré. New York: John Wiley & Sons, Inc., 1953. See pp. 102, 104, and 198.

ELEMENTARY

THREE-DIMENSIONAL WAVE FUNCTIONS

IN SPHERICAL COORDINATES

8

IN THE PREVIOUS CHAPTERS we restricted our attention to problems in one dimension, because of their mathematical simplicity. Relatively few problems can be treated realistically with this simplification, however, and to understand and be able to solve most of the basic and even elementary problems in atomic and nuclear physics it is necessary to put attention to the analysis of problems in three dimensions. In doing so we will immediately encounter differential equations different from any we have met previously, with solutions in terms of new functions which we must derive. The solutions for the fundamental bound particle problems treated in this chapter (the rigid rotator, the spherically symmetric square well, the harmonic oscillator, and Coulomb potentials) are simple polynomials, which we will derive in detail because of their recurring importance in the study of many other more complex problems. Indeed, these polynomial functions are the basic wave functions for the analysis of the more complicated problems.

8.1 GENERAL METHOD FOR SPHERICALLY SYMMETRIC POTENTIALS

A great many problems in physics, having to do with the atom, the nucleus, or nuclear interactions, for example, may be calculated exactly by a spherically symmetric potential, that is, by a potential function $V(r)$ which depends only on r and does not involve any angle variables. For this important class of problems, the three-dimensional time-independent wave function may be separated into three product wave functions,

$$\psi(r,\theta,\varphi) = R(r)\Theta(\theta)\Phi(\varphi) \tag{8.1}$$

The time-independent Schroedinger wave equation in three dimensions is in this case

$$\nabla^2\psi + \frac{2m}{\hbar^2}[E - V(r)]\psi = 0 \tag{8.2}$$

where ∇^2, the Laplacian operator in spherical coordinates, is given by

$$\nabla^2\psi = \frac{1}{r^2}\frac{\partial}{\partial r}\left(r^2\frac{\partial\psi}{\partial r}\right) + \frac{1}{r^2 \sin\theta}\frac{\partial}{\partial\theta}\left(\sin\theta\frac{\partial\psi}{\partial\theta}\right) + \frac{1}{r^2 \sin^2\theta}\frac{\partial^2\psi}{\partial\varphi^2} \tag{8.3}$$

We proceed as in Chapter 1, treating the radiation in a rectangular box, and substitute Eqs. (8.3, 8.1) into Eq. (8.2) and multiply (8.2) by $1/\psi$, to find

$$\frac{1}{R}\frac{1}{r^2}\frac{\partial}{\partial r}\left(r^2\frac{\partial R}{\partial r}\right) + \frac{1}{\Theta r^2 \sin\theta}\frac{\partial}{\partial\theta}\left(\sin\theta\frac{\partial\Theta}{\partial\theta}\right)$$

$$+ \frac{1}{\Phi}\frac{1}{r^2 \sin^2\theta}\frac{\partial^2\Phi}{\partial\varphi^2} + \frac{2m}{\hbar^2}[E - V(r)] = 0 \tag{8.4}$$

On multiplying all terms in Eq. (8.4) by $r^2 \sin^2\theta$ and placing the φ-dependent term on the right, we find that while the left-hand side contains no φ dependence, the right-hand side contains no r or θ dependence, and hence

$$\frac{1}{\Phi}\frac{\partial^2\Phi}{\partial\varphi^2} = -m^2 \tag{8.5}$$

where m is an arbitrary constant. (For reasons which will appear later, m is the magnetic quantum number; it must not be confused with the mass of a particle, also written m.)

The solution of Eq. (8.5) may be recognized to be

$$\Phi = Ae^{im\varphi} + Be^{-im\varphi} \tag{8.6}$$

In order for Φ to be single-valued (i.e., to have the same value at $\varphi + 2\pi$ as at φ), m must be an integer. By substituting Eq. (8.5) into Eq. (8.4) and multiplying (8.4) by r^2 we obtain

$$\frac{1}{R}\frac{\partial}{\partial r}\left(r^2\frac{\partial R}{\partial r}\right) + \frac{2m}{\hbar^2}r^2[E - V(r)]$$

$$= -\frac{1}{\Theta \sin \theta}\frac{\partial}{\partial \theta}\left(\sin \theta \frac{\partial \Theta}{\partial \theta}\right) + \frac{m^2}{\sin^2\theta} = K \tag{8.7}$$

where K, like m, is an arbitrary constant independent of r and θ.

Note that $\Theta(\theta)$ does *not* depend on the sign but only on the magnitude or absolute value of m. Letting $x = \cos \theta$ and noting that

$$\frac{\partial}{\partial \theta} = -\sqrt{1 - x^2}\frac{\partial}{\partial x}$$

we find

$$\frac{\partial}{\partial x}\left[(1 - x^2)\frac{\partial \Theta}{\partial x}\right] + \left(K - \frac{m^2}{1 - x^2}\right)\Theta = 0 \tag{8.8}$$

Let

$$\Theta = (1 - x^2)^{|m|/2}F(x) \tag{8.9}$$

then

$$\frac{\partial \Theta}{\partial x} = -|m|x(1 - x^2)^{(|m|/2-1)}F(x) + (1 - x^2)^{|m|/2}\frac{\partial F}{\partial x}$$

$$\frac{\partial}{\partial x}\left[(1 - x^2)\frac{\partial \Theta}{\partial x}\right] = (m^2x^2 - |m| + |m|x^2)(1 - x^2)^{(|m|/2-1)}F(x)$$

$$- 2x(|m| + 1)(1 - x^2)^{|m|/2}\frac{\partial F}{\partial x} + (1 - x^2)^{(|m|/2+1)}\frac{\partial^2 F}{\partial x^2}$$

Making this substitution into Eq. (8.8) and dividing the result by

$$(1 - x^2)^{|m|/2}$$

we obtain a differential equation for F:

$$(1 - x^2)\frac{d^2F}{dx^2} - 2(|m| + 1)x\frac{dF}{dx} + [K - |m|(|m| + 1)]F = 0 \qquad (8.10)$$

To find a satisfactory solution of Eq. (8.10) we try expanding $F(x)$ in a series of ascending powers of x,

$$F(x) = \sum_{i=0}^{\infty} a_i x^i \qquad (8.11)$$

and substitute Eq. (8.11) into Eq. (8.10) to obtain

$$(1 - x^2) \sum_{i=0}^{\infty} i(i - 1)a_i x^{i-2} - 2(|m| + 1) \sum_{i=0}^{\infty} ia_i x^i$$

$$+ [K - |m|(|m| + 1)] \sum_{i=0}^{\infty} a_i x^i = 0 \qquad (8.12)$$

In order for Eq. (8.12) to hold for any arbitrary value of x it must hold separately for each power of x in the infinite summations. That is, for the coefficients of a particular power of x, say the jth power, we find

$$(j + 2)(j + 1)a_{j+2}x^j - j(j - 1)a_j x^j$$
$$- 2(|m| + 1)ja_j x^j + [K - |m|(|m| + 1)]a_j x^j = 0$$

Thus

$$a_{j+2} = \frac{(j + |m|)(j + |m| + 1) - K}{(j + 2)(j + 1)} a_j \qquad (8.13)$$

For the start of the series where $K > (j + |m|)(j + |m| + 1)$ the coefficients of x^j in Eq. (8.11) alternate in sign. We now apply a boundary condition that the solution must satisfy, namely, that the solution must be finite over all possible values the variables may assume. The higher terms in the series all have the same sign, and for $x = \pm 1$ the series diverges unless it can be terminated. Therefore the resulting wave function can be finite only for a finite series, and this can be obtained only if for some highest value of j

$$K = (j + |m|)(j + |m| + 1) \qquad (8.14a)$$

or by letting $j + |m| = l$, whence the condition on K yielding a finite series is

$$K = l(l + 1) \qquad (8.14b)$$

where l is of course a positive integer since j and $|m|$ are positive integers. Note that the series expansion really consists of two series, an even one starting off with a_0 and an odd series starting with a_1. Both series build up in steps of two to a_l where l is an even integer for the a_0 series and odd for the a_1 series, a_{l+2} being expressible in terms of a_l. These two series may not be terminated simultaneously by the same value of K, and *therefore a solution consists of only one of the two series and contains only even powers of* $\cos \theta$ *or only odd powers but not both.* The angular wave functions, Eqs. (8.6) and (8.9), are called spherical harmonic functions.

Referring back to Eq. (8.7), we find for the radial wave equation

$$\frac{1}{r^2} \frac{\partial}{\partial r}\left(r^2 \frac{\partial R}{\partial r}\right) + \frac{2m}{\hbar^2}\left[E - V(r) - \frac{\hbar^2}{2m} \frac{l(l+1)}{r^2}\right]R = 0 \qquad (8.15)$$

Note that $R(r)$ depends on l but not on m.

8.2 ANGULAR MOMENTUM

The angular momentum of an isolated physical system is conserved. That is, like the energy it is a constant of the motion of an isolated system. Hence the eigenvalues of the angular momentum operators are very important quantities which may be used to identify a state. This is especially true for those cases when there is more than one state having the same energy and the Hamiltonian or energy operator alone is not sufficient to completely specify a state in terms of its eigenvalue, the energy. Moreover, the angular momentum operators can be used to separate that part of the particle motion which is rotational (and as we shall see, is the same for all spherically symmetric potentials) from the rest of the motion which differs for each of the many dissimilar central force problems occurring in nature. Thus, theorems on angular momentum have general validity and can be applied to all spherically symmetric systems independent of what the particular radial dependence of the potential is.

In this section we shall show that the operator for the total angular momentum is given by the angular derivative terms in the Laplacian operator Eq. (8.3) (for which we found the eigenvalue was some multiple of $l(l + 1)$). The letter l is, therefore, the quantum number for the total angular momentum. As we have seen, however, a state with given l can have different values of m. This corresponds to the fact that the eigenvalue l specifies the magnitude of the angular momentum but not its direction. As we will soon see, m represents the projection of the total angular momentum on a given axis. The letter m is called the magnetic quantum number because if it represents the angular momentum of an electric charge about a given axis, the system will have a magnetic moment about that axis. The states with different m but the same l may then be distinguished (the degeneracy in m is "broken up") by turning on a magnetic field and observing the different energies of the different states in the magnetic field.

As a first step to understanding the crucial significance of angular momentum to a theoretical interpretation of three-dimensional systems, we will express the angular momentum operator and its axial components L_z, L_x, and L_y in spherical coordinates.

The total angular momentum is defined as

$$\mathbf{L} = \mathbf{r} \times \mathbf{p} \tag{8.16}$$

The total angular momentum *operator* expressed in coordinate space is

$$\mathbf{L} = \frac{\hbar}{i}\mathbf{r} \times \nabla \tag{8.17}$$

Its z component is

$$L_z = \frac{\hbar}{i}\left(x\frac{\partial}{\partial y} - y\frac{\partial}{\partial x}\right) \tag{8.18}$$

This expression may be transformed to spherical coordinates by noting that

$$r = \sqrt{x^2 + y^2 + z^2}, \quad \tan \theta = \sqrt{(x^2 + y^2)}/z, \quad \tan \varphi = y/x$$

and

$$\frac{\partial}{\partial x} = \frac{\partial r}{\partial x}\frac{\partial}{\partial r} + \frac{\partial \theta}{\partial x}\frac{\partial}{\partial \theta} + \frac{\partial \varphi}{\partial x}\frac{\partial}{\partial \varphi}$$

$$= \frac{x}{r}\frac{\partial}{\partial r} + \frac{xz}{r^2\sqrt{x^2 + y^2}}\frac{\partial}{\partial \theta} - \frac{y}{x^2 + y^2}\frac{\partial}{\partial \varphi} \tag{8.19}$$

$$\frac{\partial}{\partial y} = \frac{y}{r}\frac{\partial}{\partial r} + \frac{yz}{r^2\sqrt{x^2 + y^2}}\frac{\partial}{\partial \theta} + \frac{x}{x^2 + y^2}\frac{\partial}{\partial \varphi} \tag{8.20}$$

Hence Eq. (8.18) becomes

$$L_z = \frac{\hbar}{i}\frac{x^2}{x^2 + y^2}\left(1 + \frac{y^2}{x^2}\right)\frac{\partial}{\partial \varphi} = \frac{\hbar}{i}\frac{\partial}{\partial \varphi} \tag{8.21}$$

The other components of the total angular momentum operator may be found in spherical coordinates in a similar way. Since

$$\frac{\partial}{\partial z} = \frac{z}{r}\frac{\partial}{\partial r} - \frac{\sqrt{x^2 + y^2}}{r^2}\frac{\partial}{\partial \theta} \tag{8.22}$$

$$L_x = \frac{\hbar}{i}\left(y\frac{\partial}{\partial z} - z\frac{\partial}{\partial y}\right) = \frac{\hbar}{i}\left(\frac{-y}{\sqrt{x^2 + y^2}}\frac{\partial}{\partial \theta} - \frac{xz}{x^2 + y^2}\frac{\partial}{\partial \varphi}\right)$$

$$= \frac{\hbar}{i}\left(-\sin\varphi\frac{\partial}{\partial \theta} - \cot\theta\cos\varphi\frac{\partial}{\partial \varphi}\right) \tag{8.23}$$

$$L_y = \frac{\hbar}{i}\left(z\frac{\partial}{\partial x} - x\frac{\partial}{\partial z}\right) = \frac{\hbar}{i}\left(\frac{x}{\sqrt{x^2 + y^2}}\frac{\lambda}{\partial \theta} - \frac{\lambda}{x^2 + y^2}\frac{\partial}{\partial \varphi}\right)$$

$$= \frac{\hbar}{i}\left(\cos\varphi\frac{\partial}{\partial \theta} - \sin\varphi\cot\theta\frac{\partial}{\partial \varphi}\right) \tag{8.24}$$

From the operators representing L_x, L_y, and L_z we may now readily find the operator for the total magnitude of the angular momentum, or rather the operator representing the square of the total angular momentum:

$$L_z^2 = -\hbar^2 \frac{\partial^2}{\partial \varphi^2}$$

$$L_x^2 = -\hbar^2 \left(\sin^2 \varphi \frac{\partial^2}{\partial \theta^2} + \cot^2 \theta \cos^2 \varphi \frac{\partial^2}{\partial \varphi^2} + 2 \sin \varphi \cos \varphi \cot \theta \frac{\partial^2}{\partial \theta \partial \varphi} \right.$$

$$\left. - \cot^2 \theta \cos \varphi \sin \varphi \frac{\partial}{\partial \varphi} - \csc^2 \theta \sin \varphi \cos \varphi \frac{\partial}{\partial \varphi} + \cot \theta \cos^2 \varphi \frac{\partial}{\partial \theta} \right)$$

$$L_y^2 = -\hbar^2 \left(\cos^2 \varphi \frac{\partial^2}{\partial \theta^2} + \sin \varphi \cot^2 \theta \frac{\partial^2}{\partial \varphi^2} - 2 \sin \varphi \cos \varphi \cot \theta \frac{\partial^2}{\partial \theta \partial \varphi} \right.$$

$$\left. + \cot^2 \theta \cos \varphi \sin \varphi \frac{\partial}{\partial \varphi} + \csc^2 \theta \sin \varphi \cos \varphi \frac{\partial}{\partial \varphi} + \sin^2 \varphi \cot \theta \frac{\partial}{\partial \theta} \right)$$

$$L^2 = L_x^2 + L_y^2 + L_z^2 = \hbar^2 \left(\frac{\partial^2}{\partial \theta^2} + \cot \theta \frac{\partial}{\partial \theta} + \csc^2 \theta \frac{\partial^2}{\partial \varphi^2} \right) \qquad (8.25)$$

or, by rearranging the derivatives in this expression,

$$L^2 = -\hbar^2 \left[\frac{1}{\sin \theta} \frac{\partial}{\partial \theta} \left(\sin \theta \frac{\partial}{\partial \theta} \right) + \frac{1}{\sin^2 \theta} \frac{\partial^2}{\partial \varphi^2} \right] \qquad (8.25a)$$

Equation (8.25a) comprises the angular derivative expressions in the Laplacian operator ∇^2:

$$r^2 \nabla^2 = \frac{\partial}{\partial r} \left(r^2 \frac{\partial}{\partial r} \right) + \frac{1}{\sin \theta} \frac{\partial}{\partial \theta} \left(\sin \theta \frac{\partial}{\partial \theta} \right) + \frac{1}{\sin^2 \theta} \frac{\partial^2}{\partial \varphi^2}$$

When spherical harmonic wave functions are used, the operator for the z component of the angular momentum, given in Eq. (8.21), has the eigenvalue $\pm m$ since the φ-dependent part of the wave function is given by one of the terms in Eq. (8.6). The square of the total angular momentum has the operator given by Eq. (8.25a), which from Eq. (8.7) and (8.14b) was found to have the eigenvalue $\hbar^2 l(l + 1)$. That is, when spherical harmonic wave functions are used,

$$L^2 \psi = \hbar^2 l(l + 1)\psi, \qquad L_z \psi = \pm m \hbar \psi \qquad (8.26)$$

Thus aside from the energy eigenvalue E, the system also has angular momentum and angular momentum axial projection eigenvalues, l and m. Further, *the eigenvalue $\hbar^2 l(l + 1)$ corresponds to the square of the total angular momentum of the system.* Classically, the rotational energy of a point mass at radius r is given in terms of the angular velocity, ω, and moment of inertia, I, by

$$\tfrac{1}{2} I \omega^2 = \frac{I^2 \omega^2}{2mr^2} \rightarrow \frac{\hbar^2 l(l + 1)}{2mr^2}$$

which is the rotational energy of a rigid rotator, to be discussed in the next section.

The angular motion about the axis from which θ is measured, the rotational axis, is given by the dependence of the wave function on φ. If, for example, j were zero in Eq. (8.14a) $|m|$ would be equal to l and the magnitude of the total angular momentum would be $\hbar \sqrt{|m|(|m| + 1)}$. In general, m is the projection of the total angular momentum vector on this polar axis (see Figure 8.1). The angular momentum about this axis, p_φ, is given by

$$p_\varphi \psi = \frac{\hbar}{i} \frac{\partial \psi}{\partial \varphi} = \hbar m \psi \qquad (8.26a)$$

Note that the eigenvalue of the operator for the angular momentum projection on the polar axis is m, not $\pm \sqrt{|m|(|m| + 1)}$ which, in general, is larger than m. (Note that we cannot have the total angular momentum vector absolutely parallel to the polar axis and satisfy the uncertainty principle since then p_x, p_y, x, and $y = 0$. m is called the magnetic quantum number since the polar axis is frequently chosen parallel to the magnetic field in problems where an external magnetic field is applied, as in the Zeeman effect. l is called the azimuthal quantum number. Since m is integral, the projection of the angular momentum vector on the polar axis is quantized; for a given l, m may take on all $2l + 1$ different values from $-l$ to $+l$. All of these states will have the same energy, and hence are called *degenerate*, unless a magnetic field is present in which case the energies will differ. In a magnetic field the electrons or nuclei will line up their angular momenta parallel or antiparallel to the magnetic field and their energy will be larger or smaller depending on their alignment in the magnetic field.

When $m = 0$, $\Theta(\theta) = F(x)$ and the series defined by Eq. (8.11) is comprised of Legendre polynomials, the number of terms in each polynomial depending on the highest possible value of i in Eq. (8.11), which is l by the condition in Eq. (8.14) for the termination of the series. When $m \neq 0$,

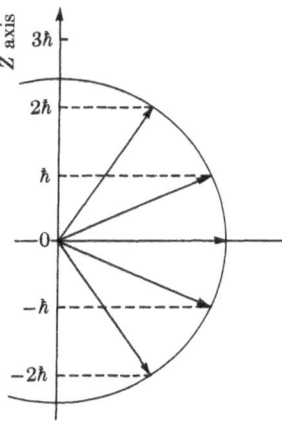

Figure 8.1   *Illustration of the quantized projection of the angular momentum vector $l\hbar$ on the Z axis for an example for which $l = 2$. Note that the length of the vector is $\hbar\sqrt{l(l + 1)} = \sqrt{6}\hbar = 2.45\hbar$ which is greater than the maximum value m may have in this case and remain integral.*

$\Theta(\theta)$ is called the associated Legendre function and the product $\Theta(\theta)\Phi(\varphi)$ is frequently expressed as $Y_l{}^m(\theta,\varphi)$ and called a surface harmonic. A very lucid discussion and detailed table of these functions is given in Chapter 5 of Pauling and Wilson (see Bibliography of this chapter). If these functions are normalized, that is

$$\int_0^{2\pi} \int_0^\pi \left| Y_l{}^m(\theta,\varphi) \right|^2 \sin \theta \, d\theta \, d\varphi = 1$$

then they are given by

$$Y_l^m(\theta,\varphi) = \frac{1}{\sqrt{2\pi}} e^{\pm im\varphi}\Theta_l^{|m|}(\theta)$$

$$\Theta_l^{|m|}(\theta) = (-1)^l \sqrt{\frac{(2l + 1)(l + |m|)!}{2(l - |m|)!}} \frac{1}{2^l l!} \frac{1}{\sin^{|m|}(\theta)} \left(\frac{d}{d(\cos\theta)}\right)^{l-|m|} \sin^{2l}(\theta)$$

Graphs of the first few associated Legendre polynomials are illustrated in Figures 8.2 and 8.3. A short table is presented in Section 8.4. Note from Table 8.1 and Figures 8.2 and 8.3 that the associated Legendre functions

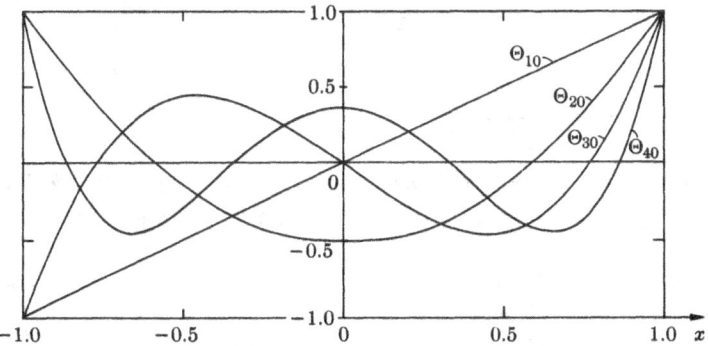

Figure 8.2 *Zero-order* $(m = 0)$ *associated Legendre functions* $\Theta_{lm}(x)$ *up to the fourth degree* $(l = 4)$. $x = \cos\theta$. *Note that the odd-degree functions are odd functions of x while the even-degree functions are even functions of x.* (*From Jahnke, Emde, and Lösch,* Tables of Higher Functions, *Fig. 69. Stuttgart: B. G. Teubner Verlagsgesellschaft, 1960*).

are even or odd functions about $\theta = \pi/2$ depending on whether $(m + l)$ is even or odd. Since also $e^{im\varphi}$ is even or odd about $\varphi = \pi$ ($e^{im\Omega+\pi} = (-1)^m e^{im\varphi}$) depending on whether m is even or odd, the surface harmonic $Y_l^m(\Theta, \varphi)$, is even or odd over the range of the θ and φ variables depending on whether $(m + l + m)$ is even or odd. Therefore, since 2^m is always even, the surface harmonics are even or odd functions (with respect to changing θ to $\pi - \theta$

and φ to $\varphi + \pi$ which corresponds to changing $\mathbf{r}$ to $-\mathbf{r}$) depending on whether l is even or odd.

The angular wave function $\Theta(\theta)$ is spherically symmetric about the origin when $l = 0$; when $m = l$, Θ is increasingly concentrated in the xy plane as l increases. When l is constant Θ is increasingly concentrated in

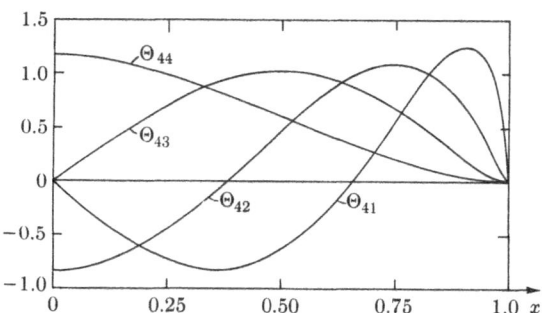

Figure 8.3   *Associated Legendre functions, $\Theta_{4m}(x)$, of fourth degree and order m for $m = 1, 2, 3, 4$. $x = \cos\theta$. Only values for positive x are shown ($0 \leqslant \theta \leqslant \pi/2$), but as may be surmised from the figure, fourth degree functions of odd order are odd functions of x. (Functions of odd degree and even order are also odd, while odd-degree, odd-order and even-degree, even-order functions are even.) (From Jahnke, Emde, and Lösch, op. cit., Figs. 72–75.)*

the xy plane for the case $l = 3$, as illustrated in Figure 8.4. Note that in conformity with the uncertainty principle the particle motion is not entirely confined to the xy plane even for $m = l$. For this reason the angular momentum projected on the z axis must always be less than the total angular momentum.

8.3 ANGULAR MOMENTUM COMMUTATION RELATIONS

We conclude our discussion of the angular momentum by examining a few very useful and important rules for commutation of the total angular

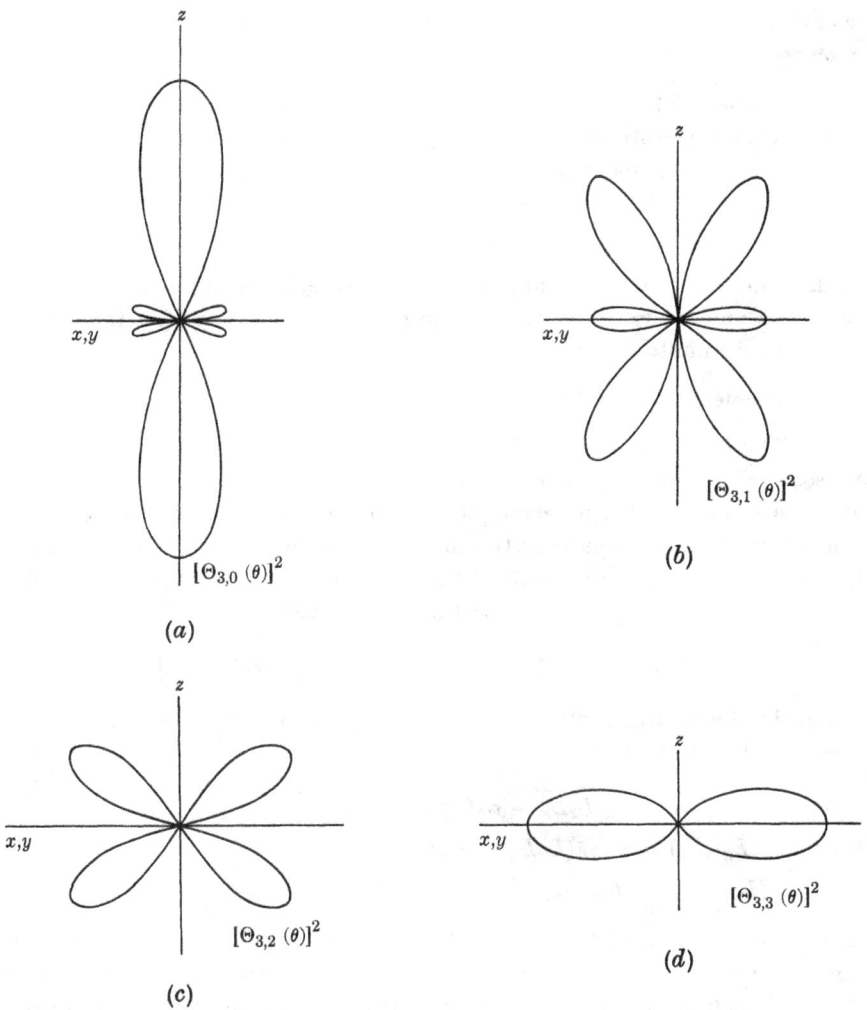

Figure 8.4 *Polar graphs of the probability of finding a particle at θ if its wave function is* $\Theta_{3m}(\theta)$ *($l = 3$) for m = 0, 1, 2, and 3. (The projection of the total angular momentum vector on the z axis is 0, 1, 2, and 3 respectively in the four cases.)*

momentum, its components, and operators representing position and linear momentum.

Problem 8.1   Prove that $[L_x, L_y] = i\hbar L_z$, where $[A, B]$ means the commutation operation of the two operators A and B; that is $[A, B] \equiv AB - BA$. Prove also that $[L_y, L_z] = i\hbar L_x$, and $[L_z, L_x] = i\hbar L_y$, that is, prove that in general $\mathbf{L} \times \mathbf{L} = i\hbar \mathbf{L}$.

The significance of the commutation properties of two operators is that the two physical quantities they represent may be simultaneously measured and their eigenvalues used to label the same wave function, if the operators commute.

Problem 8.2   Find $[L_z, x]$, $[L_z, y]$ and $[L_z, p_x]$.

The components of $\mathbf{L}$ do not commute with one another, and thus we see that only one component of the angular momentum can be measured at a time (L_x and L_y, for example, cannot be measured simultaneously). The operator for the square of the total angular momentum does commute, however, with any component of the angular momentum. Thus, L_z, for example, and L^2 may be measured simultaneously for a given system:

$$[L^2, L_z] = L_x^2 L_z - L_z L_x^2 + [L_y^2, L_z] + [L_z^2, L_z]$$

Using the theorem of Problem 8.1 the first three terms on the right may be shown to add up to zero

$$L_x^2 L_z = L_x(L_z L_x - i\hbar L_y) = L_z L_x^2 - i\hbar L_y L_x - i\hbar L_x L_y$$

thus $\quad [L_x^2, L_z] = -i\hbar(L_x L_y + L_y L_x)$

$$[L_y^2, L_z] = i\hbar(L_x L_y + L_y L_x)$$

Also, L_z commutes with L_z^2, and thus L^2 is proved to commute with L_z. Similarly, L_y and L_x commute with L^2, or in general, $\mathbf{L}$ commutes with L^2. Although no two components of $\mathbf{L}$ may be measured simultaneously, we know that when L_z is a maximum (that is, when $\langle L_z \rangle = m = l$), $\langle L_x \rangle = \langle L_y \rangle = 0$; in this case, however, L_x and L_y are not completely determined since

$$\langle L_x^2 \rangle = \langle L_y^2 \rangle = \tfrac{1}{2}\langle L^2 - L_z^2 \rangle = \tfrac{1}{2}[l(l+1) - m^2]\hbar^2 = \tfrac{1}{2}l\hbar^2 \quad (8.27)$$

Problem 8.3 What is the possible result of a measurement of L_z and L^2 on the state represented by $\psi = R(r) \sin \theta \cos \varphi$?

8.4 THE RIGID ROTATOR

We shall treat rigid rotation which means that the radial position of every point is fixed, but much of the discussion is also valid for many problems in spherical coordinates. In the case of rigid rotation all the energy is rotational and

$$E = \frac{\hbar^2}{2m} \frac{l(l+1)}{r^2} \tag{8.28}$$

The rigid rotator wave functions are simply the angle-dependent wave functions, i.e., the spherical harmonic wave functions discussed in the previous section:

$$\psi = Y_l{}^m(\theta, \varphi) = \Theta_l{}^m(\theta)\Phi(\varphi)$$

A few of these functions are listed in Table 8.1.

Table 8.1 *The associated Legendre functions of the first kind, $\Theta_{lm}(\theta)$, up to $l = m = 3$. The functions for $m = 0$ are proportional to the ordinary Legendre functions. Note that the evenness or oddness of the functions about $\theta = \pi/2$ depends on the evenness or oddness of $(m + l)$.*

$$\Theta_{00} = \frac{\sqrt{2l+1}}{\sqrt{2}} = \frac{1}{\sqrt{2}}$$

$$\Theta_{10}(\theta) = \sqrt{\tfrac{3}{2}} \cos \theta \qquad\qquad \Theta_{11}(\theta) = \frac{\sqrt{3}}{2} \sin \theta$$

$$\Theta_{20}(\theta) = \sqrt{\tfrac{5}{2}}(\tfrac{3}{2} \cos^2 \theta - \tfrac{1}{2}) \qquad\qquad \Theta_{21}(\theta) = \frac{\sqrt{15}}{2} \sin \theta \cos \theta,$$

$$\Theta_{22}(\theta) = \frac{\sqrt{15}}{4} \sin^2 \theta$$

$$\Theta_{30}(\theta) = \tfrac{3}{4}\sqrt{14} \cos \theta \, (\cos^2 \theta - 1) \qquad\qquad \Theta_{31}(\theta) = \frac{\sqrt{42}}{8} \sin \theta \, (5 \cos^2 \theta - 1)$$

$$\Theta_{32}(\theta) = \frac{\sqrt{105}}{4} \sin^2 \theta \cos \theta \qquad\qquad \Theta_{33}(\theta) = \frac{\sqrt{70}}{8} \sin^3 \theta$$

When a rotating molecular system absorbs or emits photons, the photons will have a frequency spectrum in the overwhelmingly most probable case where l changes by one unit (discussed later in Section 10.4) given by

$$\hbar\omega = \frac{\hbar^2}{2I}[(l + 2)(l + 1) - (l + 1)l]$$

$$= \frac{\hbar^2}{2I}2(l + 1)$$

where the spectrum frequencies are proportional to the integer series 1, 2, 3, 4, ... The pre-wave-theory Bohr–Sommerfeld quantum rules (which we have seen to result from the first-order WKB approximation to the wave theory) predicts a frequency spectrum of

$$\hbar\omega = \frac{\hbar^2}{2I}[(l + 1)^2 - l^2] = \frac{\hbar^2}{2I}(2l + 1) \propto 1, 3, 5, \ldots$$

Rotational transitions are frequently coupled with vibrational transitions so that when a vibrational transition as well as a rotational transition occurs radiation may be emitted and energy conserved if l either decreases or increases. Thus the emitted or absorbed photons have a frequency spectrum twice as broad as for pure rotational transitions in which l can only decrease. Hence the spectrum is proportional to ± 1, ± 2, ± 3, $\pm 4 \ldots$ on the basis of the wave theory or, ± 1, ± 3, $\pm 5 \ldots$ on the basis of the Bohr–Sommerfeld quantum conditions. Thus the wave theory predicts evenly spaced frequencies with a double space in the center between $+1$ and -1 whereas the pre-wave theory predicts evenly spaced lines everywhere. The observed spectrum illustrated in Figure 8.5 consists of evenly spaced lines with a gap at the center which was once a big puzzle, as the Bohr–Sommerfeld quantum conditions did not predict the experimentally observed gap.

Problem 8.4   Represent the voltage on a sphere by a series of spherical harmonics up to the fourth term for voltages on the two hemispheres of 100 volts for $0 < \theta < \pi/2$ and 0 volts for $\pi/2 < \theta < \pi$. (The coefficients of the spherical harmonics which are orthogonal functions are obtained as for the Fourier series coefficients in Chapter 2.

$$A_l = \int_0^\pi f(\theta) \Theta_l{}^0(\theta) \sin \theta \, d\theta$$

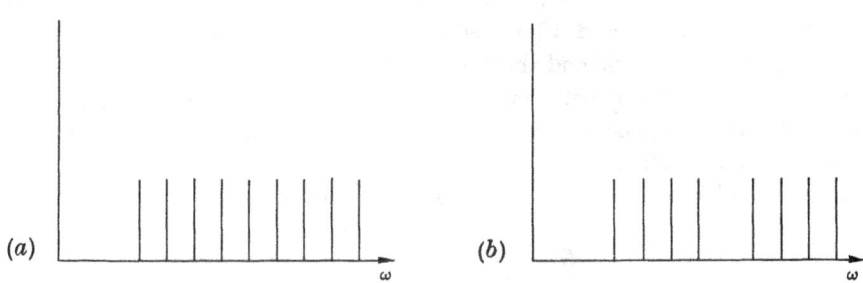

Figure 8.5 (a) *Illustration of rotational–vibrational spectrum predicted by the Bohr–Sommerfeld quantum conditions.* (b) *Illustration of rotational–vibrational spectrum observed and predicted by quantum mechanics.*

Problem 8.5 Find the energy levels of the HCl molecule at temperatures too low for vibration to be important, assuming that only rotation occurs. The moment of inertia in the appropriate coordinate system (center of mass) is given by $I = [m_H m_{Cl}/(m_H + m_{Cl})]r_0{}^2$ where r_0 is 1.27 Å.

Problem 8.6 The voltage on the surface of a sphere is 100 volts on one octant and 0 elsewhere. Express the voltage approximately as a sum of spherical harmonics with $l \leqslant 2$.

Problem 8.7 Express the function $x^2 - 1$ as a series of Legendre polynomials for $-1 \leqslant x \leqslant 1$.

8.5 THE SQUARE WELL

For many potentials a change is frequently made in the dependent variable in Eq. (8.15) by letting

$$R(r) = \frac{u(r)}{r}$$

Thus we obtain

$$\frac{\partial^2 u}{\partial r^2} + \frac{2m}{\hbar^2}\left[E - V(r) - \frac{\hbar^2}{2m}\frac{l(l+1)}{r^2}\right]u = 0 \qquad (8.29)$$

The reader will readily recognize Eq. (8.29) as the one-dimensional wave equation in r which was used to treat the deuteron in Section 6.2 with $l = 0$ for the ground state and $V(r)$ a square well. In general, if the potential is spherically symmetric and there is no rotation (that is, when l is zero), the substitution of u for R will change the radial equation into one resembling a simple one-dimensional problem in Cartesian coordinates. When l is not zero, the problem may be handled by treating the term in $l(l + 1)$ as an additional potential function. The effective potential is then given by

$$V'(r) = V(r) + \frac{\hbar^2}{2m}\frac{l(l+1)}{r^2} \qquad (8.30)$$

The term in l which represents the rotational energy is frequently referred to as the centrifugal potential or centrifugal barrier since it acts as the potential of a repulsive force, tending to prevent particles of high angular momentum (that is, large l) from approaching very close to the center of force ($r = 0$). In classical physics, only particles with energy greater than

$$\frac{\hbar^2}{2m}\frac{l(l+1)}{r_1^2}$$

may approach to within $r = r_1$ if energy is to be conserved. In quantum mechanics the particle wave function falls off as r^l in regions where the centrifugal potential exceeds the particle energy, thus decreasing but not eliminating the probability of finding the particle at small r. The relevant energy diagram is shown in Figure 8.6.

For a square well potential of depth $-V_0$ and arbitrary l, Eq. (8.15) becomes for $r \leqslant a$

$$\frac{\partial^2 R}{\partial r^2} + \frac{2}{r}\frac{\partial R}{\partial r} + \left(\frac{p_1^2}{\hbar^2} - \frac{l(l+1)}{r^2}\right)R = 0 \qquad (8.31)$$

where $p_1^2 = 2m(E + V_0)$. Let $\rho = (p_1/\hbar)r$ and $R = F/\sqrt{\rho}$: then

$$\frac{\partial^2 F}{\partial \rho^2} + \frac{1}{\rho}\frac{\partial F}{\partial \rho} + \left(1 - \frac{(l + \frac{1}{2})^2}{\rho^2}\right)F = 0$$

which is Bessel's differential equation of order $l + \frac{1}{2}$. Thus for $r \leqslant a$ the radial wave function is

$$R = Ar^{-1/2}J_{l+\frac{1}{2}}(p_1r/\hbar) + Br^{-1/2}J_{-(l+\frac{1}{2})}(p_1r/\hbar) \qquad (8.32)$$

The solutions are spherical Bessel and Neumann functions, the latter being essentially Bessel functions of negative order (see Morse and Feshbach, and Schiff, cited in Bibliography of this chapter). A graph of a few of these

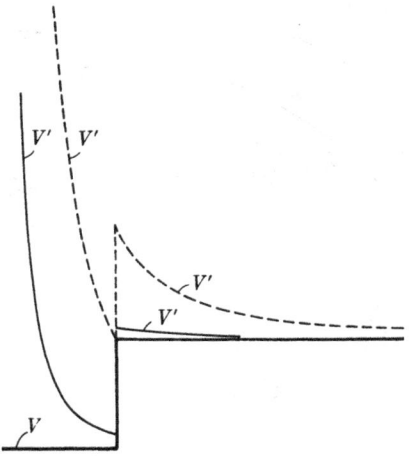

Figure 8.6   *Energy graphs of a square well potential $V(r)$ and the effective potential $V'(r)$ which includes the centrifugal barrier in addition to the square well for two different cases, one in which bound states are possible (——) and the other for larger l where V' is everywhere positive and no bound states are possible (– –).*

functions is presented in Figure 8.7. (The Neumann function is not, in general, a Bessel function of negative order; this is true only for half-integral order.)

For $r \geqslant a$ and negative E, p^2 is negative and spherical Bessel and Neumann functions of imaginary argument result, which rearranged are called spherical Hankel functions which depend exponentially on r. The permissible values for the energy of the bound particle are found by requiring the wave function and its derivative to be continuous at the boundary:

$$r \leqslant a: \quad R = \frac{A}{\sqrt{r}} J_{l+\frac{1}{2}}(p_1 r/\hbar) \qquad (8.32a)$$

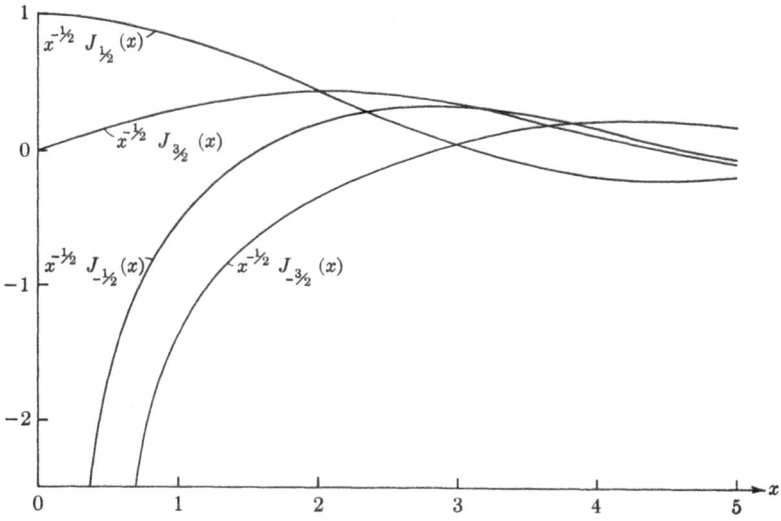

Figure 8.7 *The first two spherical Bessel and Neumann functions as function of x. Note that the Neumann functions are infinite at the origin. The behavior of all the functions begins to approach sinusoidal, with slowly diminishing amplitude, at x ~ π.*

where $p_1 = \sqrt{2m(V_0 + E)}$ and $E < 0$, and

$$r \geqslant a: \quad R = \frac{C}{\sqrt{r}} [J_{l+\frac{1}{2}}(p_2 r/\hbar) - J_{-(l+\frac{1}{2})}(p_2 r/\hbar)]$$

$$\underset{r \to \infty}{\lim} (-1)^{n+1} \sqrt{\frac{2p_2}{\hbar\pi}} \frac{C}{r} e^{-p_2 r/\hbar} \qquad (8.32b)$$

where $p_2 = \sqrt{2m(-E)}$. The Neumann function which would appear in

Eq. (8.32a) has been eliminated by requiring a finite wave function at the origin. One of the two integration constants in the solution for $r < a$ has been chosen so that we obtain a combination of $J_{l+\frac{1}{2}}$ and $J_{-(l+\frac{1}{2})}$ whose asymptotic form does not include an increasing exponential function of r as r increases without limit. Equations (8.32a,b) and their first derivatives must be equal at $r = a$; therefore R is continuous;

$$\frac{dR}{dr} \text{ is continuous}; \quad \frac{d(\log R)}{dr} = \frac{1}{R}\frac{dR}{dr} \text{ is continuous}$$

Hence

$$[(p_1/\hbar)J'_{l+\frac{1}{2}}(p_1 a/\hbar) - (2a)^{-1}J_{l+\frac{1}{2}}(p_1 a/\hbar)]/J_{l+\frac{1}{2}}(p_1 a/\hbar)$$
$$= \{(p_2/\hbar)[J'_{l+\frac{1}{2}}(p_2 a/\hbar) - J'_{-(l+\frac{1}{2})}(p_2 a/\hbar)] - (2a)^{-1}[J_{l+\frac{1}{2}}(p_2 a/\hbar) - J_{-(l+\frac{1}{2})}(p_2 a/\hbar)]\}$$
$$\times \{J_{l+\frac{1}{2}}(p_2 a/\hbar) - J_{-(l+\frac{1}{2})}(p_2 a/\hbar)\}^{-1} \qquad (8.33)$$

where the prime on the Bessel function denotes the derivative with respect to the entire argument of the Bessel function. Thus a discrete energy spectrum results, for only those values of p_1 and p_2 (and therefore only certain values of E) satisfying Eq. (8.33) are possible. For many problems the energy determination may be satisfactorily simplified by making the well infinite, in which case R must vanish identically at $r = a$. The energy spectrum is then determined by examining the nodes of the spherical Bessel functions, which are

$$J_{l+\frac{1}{2}}(p_1 a/\hbar) = 0$$

A convenient approximation for the higher eigenvalues is given by

$$J_{l+\frac{1}{2}}(p_1 r/\hbar) \xrightarrow{(p_1 r/\hbar) \gg 1} \sqrt{\frac{2\hbar}{\pi p_1 r}} \cos\left(\frac{p_1 r}{\hbar} - \frac{\pi}{2}(l+1)\right)$$

so that for $p_1 a/\hbar \gg 1$

$$\frac{p_1 a}{\hbar} = \frac{\pi}{2}[(l+1) + (2n+1)]$$

where n is an integer, and

$$E \sim -V_0 + (\hbar^2/32ma^2)(2n+l+2)^2 \qquad (8.34)$$

Note that states of either even l or odd l are degenerate in this approximate expression in that different pairs of values of l and n yield the same energy.

Problem 8.8   An electron moves in an infinite spherical square well, that is, inside a hollow sphere with impenetrable walls of radius 1 Å. Find the lowest energy level with $l = 1$. (Note that the approximate result is not valid for the lowest energy level.)

8.6   THE HYDROGEN ATOM

If the potential energy of a particle consists of the electrostatic attraction between a positive charge Ze and a negative charge $-e$, Eq. (8.15) becomes

$$\frac{1}{r^2}\frac{d}{dr}\left(r^2\frac{dR}{dr}\right) + \frac{2m}{\hbar^2}\left[E + \frac{Ze^2}{r} - \frac{\hbar^2}{2m}\frac{l(l+1)}{r^2}\right]R = 0 \qquad (8.35)$$

We now attempt to solve this equation by making a series expansion in positive powers of r, similar to the way we generated the associated Legendre functions. To simplify Eq. (8.35) let

$$n = [m(Ze^2)^2/2\hbar^2(-E)]^{1/2} = [(Ze)^2/2(-E)a_0]^{1/2} \qquad (8.36a)$$

where the constant $a_0 = \hbar^2/me^2$ is called the first Bohr radius, for historical reasons. Also, let

$$\rho = \sqrt{8m(-E)/\hbar^2}\, r \qquad (8.36b)$$

and

$$R = e^{-\rho/2}F(\rho) \qquad (8.36c)$$

We then obtain from (8.35)

$$\frac{\partial^2 F}{\partial\rho^2} + \left(\frac{2}{\rho} - 1\right)\frac{\partial F}{\partial\rho} + \left[\frac{n}{\rho} - \frac{1}{\rho} - \frac{l(l+1)}{\rho^2}\right]F = 0 \qquad (8.37)$$

where E is negative for the bound electron which this problem assumes. Equation (8.36c) is suggested by the solution of the equation to which Eq. (8.35) reduces in the limit of infinitely large r where (8.35) may be written

$$\frac{d^2R}{dr^2} + \frac{2m}{\hbar^2}ER = 0$$

Let

$$F(\rho) = \sum_{s=0}^{\infty} a_s \rho^s \qquad (8.38)$$

where the sum is taken over only positive values of s if the wave function is to be well-behaved at the origin. Substitution of Eq. (8.38) into (8.37) enables us to obtain

$$\sum_{s=0}^{\infty} \{s(s-1)a_s\rho^{s-2} + 2sa_s\rho^{s-2} - sa_s\rho^{s-1}$$
$$+ (n-1)a_s\rho^{s-1} - l(l+1)a_s\rho^{s-2}\} = 0 \qquad (8.39)$$

Since this equation must hold for every value of s it must hold for each power of s taken individually. Hence for the coefficients of a particular power ρ^{j-1} we obtain

$$j(j+1)a_{j+1} + 2(j+1)a_{j+1} - ja_j + (n-1)a_j - l(l+1)a_{j+1} = 0$$
$$(8.40)$$

from which we obtain the recursion relation in the coefficients of the series,

$$a_{j+1} = \frac{j - (n-1)}{j(j+1) + 2(j+1) - l(l+1)} a_j \qquad (8.41)$$

Since j and l are both integers, the denominator will be zero for that value of s for which $(s+1)(s+2) = l(l+1)$, that is for $s = l - 1$. Unless the coefficients of ρ^s are identically zero for $s \leqslant l - 1$, the coefficients of ρ^s in Eq. (8.38) will be infinite for $s > l - 1$. Therefore the first nonvanishing term in a series representing an acceptable wave function must be $a_l\rho^l$. On the other hand the series must be terminated, since for large s, s greater than some large integer N, $a_{s+1} \sim a_s/s \sim N!a_N/s!$, and the summation becomes approximately

$$F(\rho) = \sum_{s=l}^{N} a_s\rho^s + N!a_N \sum_{s=N}^{\infty} \frac{\rho^s}{s!}$$
$$= \sum_{s=l}^{N} a_s\rho^s + N!a_N e^\rho - \sum_{s=0}^{N} \frac{\rho^s}{s!} \qquad (8.42)$$

At large ρ the first and third terms $\Sigma \, a_s \rho^s$, $\Sigma \, \rho^s/s!$ when multiplied by $e^{-\rho/2}$ lead to an acceptable wave function, but the second term blows up too rapidly. Hence only a finite series leads to an acceptable wave function for a bound electron. The series can be terminated only by requiring that

$$n = k + 1 \qquad (8.43)$$

where k is the highest term in the series and n is defined by Eq. (8.36a). Thus

$$- E = \frac{m(Ze^2)^2}{2\hbar^2} \frac{1}{n^2} \qquad (8.44)$$

where n must be an integer from the condition expressed by Eq. (8.43). It is more convenient to redefine $F(\rho)$ as

$$F(\rho) = \rho^l \sum_{s=0}^{n-(l+1)} a_s \rho^s = \rho^l L(\rho) \qquad (8.45)$$

The finite summation over s, $L(\rho)$, is called an associated Laguerre polynomial. n is called the total quantum number and $n - (l + 1)$ the radial

Table 8.2 *The associated Laguerre functions* $L_{n-(l+1)}(\rho)$ *up to* $n - (l + 1) = l = 2$

$L_{10} = 1$	$L_{21} = 1$	$L_{32} = 1$
$L_{20} = 2 - \rho$	$L_{31} = 4 - \rho$	$L_{42} = 6 - \rho$
$L_{30} = 6 - 6\rho + \rho^2$	$L_{41} = 20 - 10\rho + \rho^2$	$L_{52} = 42 - 14\rho + \rho^2$

quantum number; $(2l + 1)$ is the order of the associated Laguerre polynomial, and $n - (l + 1)$, its highest power, is called its degree. A short table of Laguerre polynomials is presented in Table 8.2. The radial distribution of an electron in the hydrogen atom is presented in Figure (8.8) for the ground state and first few excited states. Note that the number of nodes occurring in the radial wave function is given by $n - l - 1$. (A node is defined as a point where the wave function is zero.) The normalization of the radial wave functions may be found from the requirement that

$$\int_0^{\infty} r^2 \left| R_{l,n}(r) \right|^2 dr = 1$$

Equation (8.44) is the well-known empirical Balmer formula giving the spectral lines of hydrogen. Note that for a simple Coulomb potential

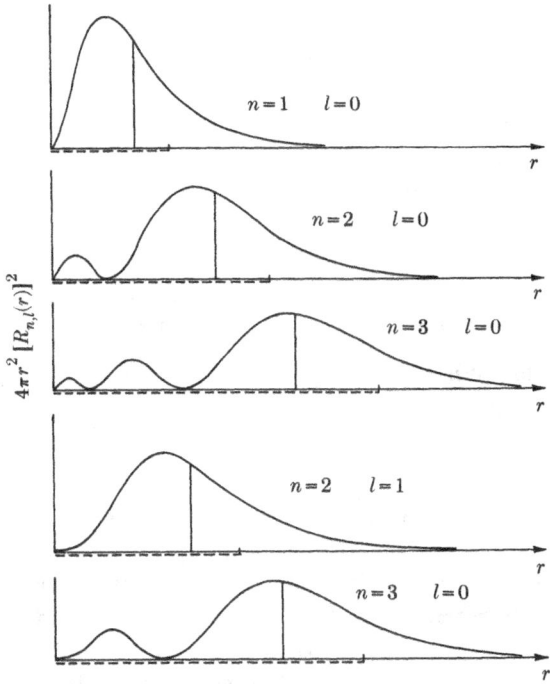

Figure 8.8 *Graph of the probability of finding an electron at radius r in a hydrogen atom for various radial wave functions R_{nl}. The dashed lines give the spread in radial position for the ground state as predicted by the old Bohr theory. The vertical lines are for the average value of r ($\langle r \rangle = \int r\psi^*\psi \, d\tau$).*

the energy depends only on the total quantum number n. For a given energy there are $2l + 1$ wave functions having the same n and l, and a total number of n^2 wave functions having the same n and giving the same value for the electron energy,

$$\sum_{l=0}^{n-1} (2l + 1) = (n - 1)n + n = n^2$$

since

$$1 + 2 + 3 + \ldots + l = \frac{l}{2}(l + 1)$$

The degeneracy in l is a special feature of the Coulomb field alone. This degeneracy is broken up by any small additional effects, for example a small magnetic interaction between the electron spin and its orbital motion. An externally applied magnetic field or the nuclear magnetic moment causes a further slight difference in energy between states of the same n and l but different m.

Problem 8.9   Draw the wave function of the hydrogen atom for $n = 4$, $l = 2$, $m = 0$. Draw $|\psi|^2 r^2$ as a function of r.

Problem 8.10   What is the expectation of r and r^2 in the ground state of the hydrogen atom? In the first excited state for which $l = 0$?

8.7 THE HARMONIC OSCILLATOR

As well as the infinite square well, the harmonic oscillator is of great interest because of the fairly recent revival of interest in the individual particle or shell model of the atomic nucleus. In this model the individual neutrons and protons move freely in a modified square well or, more commonly, a modified harmonic oscillator potential, and the harmonic oscillator energy levels and wave functions are frequently taken as a first approximation to nucleon energy levels and wave functions. Despite the similarity of the solution of this problem to that of the hydrogen atom we undertake the solution because of its great interest. The harmonic oscillator potential may be written

$$V = \tfrac{1}{2}kr^2 \qquad (8.46)$$

where k is a spring constant $= m\omega^2$ for frequency of oscillation ω.

Thus the radial wave equation for a central harmonic oscillator potential is

$$\frac{1}{r^2}\frac{\partial}{\partial r}\left(r^2\frac{\partial R}{\partial r}\right) + \frac{2m}{\hbar^2}\left[E - \tfrac{1}{2}kr^2 - \frac{\hbar^2}{2m}\frac{l(l+1)}{r^2}\right]R = 0 \qquad (8.47)$$

At large r only the oscillator potential contributes effectively to the energy, so that the wave equation becomes

$$\frac{1}{r^2}\frac{\partial}{\partial r}\left(r^2\frac{\partial R}{\partial r}\right) - \frac{mk}{\hbar^2}r^2R = 0 \qquad (8.48)$$

whose solution is

$$R = Ae^{-(\sqrt{mk}/2\hbar)r^2}$$

This form of the solution of the wave equation for large r suggests that we make the substitution

$$R = e^{-\rho^2/2}F(\rho) \qquad (8.49)$$

where

$$\rho = (\sqrt{mk}/\hbar)^{1/2}r$$

Equation (8.48) then becomes, after multiplication by $(\hbar/\sqrt{mk})e^{\rho^2/2}$,

$$\frac{\partial^2 F}{\partial \rho^2} + \left[\frac{2}{\rho} - 2\rho\right]\frac{\partial F}{\partial \rho} + \left[\frac{2}{\hbar}\sqrt{\frac{m}{k}}E - 3 - \frac{l(l+1)}{\rho^2}\right]F = 0 \qquad (8.50)$$

Let

$$F = \rho^t \sum_{t=0}^{\infty} a_t\rho^t \qquad (8.51a)$$

where t is a parameter to be determined later (t must be positive in order for the solution to be finite at the origin). If Eq. (8.51a) is substituted into Eq. (8.50) we obtain

$$\sum_{i=0}^{\infty} \left\{ (t + i)(t + i - 1)a_i\rho^{t+i-2} \right.$$

$$\left. + \left(\frac{2}{\rho} - 2\rho\right)(t + i)a_i\rho^{t+i-1} + \left(\frac{2}{\hbar}\sqrt{\frac{m}{k}}E - 3 - \frac{l(l + 1)}{\rho^2}\right)a_i\rho^{t+i} \right\} = 0$$

or if we group coefficients of the same power of ρ:

$$\{t(t - 1) + 2t - l(l + 1)\}a_0\rho^{t-2} + \{(t + 1)t + 2(t + 1) - l(l + 1)\}a_1\rho^{t-1}$$

$$+ \sum_{i=2}^{\infty} \{[(t + i)(t + i - 1) + 2(t + i) - l(l + 1)]a_i$$

$$+ [- 2(t + i - 2) + 2\sqrt{m/k}\,E/\hbar - 3]a_{i-2}\}\rho^{t+i-2} = 0$$

If Eq. (8.52) is to be true for all r $(= \rho(\sqrt{mk}/\hbar)^{-1/2})$, it must hold independently for all coefficients of the same power of r. If the coefficients of the lowest two powers of ρ are to be zero, then either

$$t(t + 1) - l(l + 1) = 0 \quad \text{and} \quad a_1 = 0$$

or

$$(t + 1)(t + 2) - l(l + 1) = 0 \quad \text{and} \quad a_0 = 0$$

In other words, either $t = l$ for a series which starts with a_0 or $t = l - 1$ for a series which starts with a_1. In order for the higher terms in Eq. (8.52) to be valid,

$$a_{i+2} = \frac{2(i + t) - 2\sqrt{m/k}\,E/\hbar + 3}{(t + i + 2)(t + i + 3) - l(l + 1)} a_i \tag{8.53}$$

The series diverges for infinite i, behaving like e^{ρ^2} at large r, so that R will not be a satisfactory wave function unless the series is terminated, as in previous problems in this chapter. Therefore, as in any bound state problem, only certain discrete values of the energy are permitted:

$$E = \hbar\sqrt{k/m}\,(j + t + \tfrac{3}{2}) = \hbar\omega(j + t + \tfrac{3}{2}) \tag{8.54a}$$

where ω^2 has been substituted for k/m and j is the highest term in the series. From Eq. (8.53) and the condition that either a_0 or a_1 must be zero, we conclude that there can only be either an even series in i starting with a_0 or an odd i series starting with a_1. *The solution cannot include both odd and even powers of ρ times ρ^t.* If $a_1 = 0$ and we let $s = i/2$ and $n = j/2$, Eq. (8.51a) and (8.54a) become

$$F_{n,l} = \rho^l \sum_{s=0}^{n} a_{2s}\rho^{2s} \tag{8.51b}$$

$$E = \hbar\omega(2n + l + \tfrac{3}{2}) \tag{8.54b}$$

where the subscripts on F are now necessary to distinguish between the various solutions or eigenfunctions having different eigenvalues n and l. If $a_0 = 0$ and we let $s = (i - 1)/2$ and $n = (j - 1)/2$, then

$$F_{n,l} = \rho^{l-1} \sum_{i=1}^{j} a_i\rho^i = \rho^l \sum_{s=0}^{n} a_{2s+1}\rho^{2s}$$
$$E = \hbar\omega(2n + 1 + l - 1 + \tfrac{3}{2}) = \hbar\omega(2n + l + \tfrac{3}{2})$$

Note that the oddness or evenness of the series is entirely dependent on l. It may be seen at once that either series (odd or even i) yields the same two formulas, Eqs. (8.51b, 8.54b).

Utilizing the fact that we need work with only one of the two series (the other being identical) and choosing even i for which $t = l$, we may rewrite Eq. (8.53) as

$$a_{t+2} = \frac{2(2n - i)}{(i + 2)(i + 2l + 3)} a_t \tag{8.53a}$$

The polynomial $F_{n,l}$ is called the spherical Hermite polynomial. Table 8.3 lists the first few spherical Hermite polynomials up to $n = l = 3$. n here is the radial quantum number. (n is usually used for the radial quantum number; however, for the hydrogen atom where a special degeneracy between the radial and orbital quantum numbers occurs, it is more convenient to work with the total quantum number instead, which is also usually labeled n.) In the case of hydrogen the total quantum number giving the energy eigenvalue is given by the sum of the radial quantum number plus the orbital quantum number.

Table 8.3 *The spherical Hermite polynomials F_{nl} (unnormalized) up to $n = l = 3$.*

$$F_{00} = 1 \qquad F_{01} = \rho \qquad F_{02} = \rho^2 \qquad F_{03} = \rho^3$$

$$F_{10} = 1 - \tfrac{2}{3}\rho^2 \qquad F_{11} = \rho(1 - \tfrac{2}{5}\rho^2) \qquad F_{12} = \rho^2(1 - \tfrac{2}{7}\rho^2) \qquad F_{13} = \rho^3(1 - \tfrac{2}{9}\rho^2)$$

$$F_{20} = 1 - \tfrac{4}{3}\rho^2 + \tfrac{4}{15}\rho^4 \qquad F_{21} = \rho \, (1 - \tfrac{4}{5}\rho^2 + \tfrac{4}{35}\rho^4)$$

$$F_{22} = \rho^2(1 - \tfrac{4}{7}\rho^2 + \tfrac{4}{63}\rho^4) \qquad F_{23} = \rho^3(1 - \tfrac{4}{9}\rho^2 + \tfrac{4}{99}\rho^4)$$

$$F_{30} = 1 - 2\rho^2 + \tfrac{4}{5}\rho^4 - \tfrac{8}{105}\rho^6 \qquad F_{31} = \rho \, (1 - \tfrac{6}{5}\rho^2 + \tfrac{12}{35}\rho^4 - \tfrac{8}{315}\rho^6)$$

$$F_{32} = \rho^2(1 - \tfrac{6}{7}\rho^2 + \tfrac{4}{21}\rho^4 - \tfrac{8}{693}\rho^6) \qquad F_{33} = \rho^3(1 - \tfrac{2}{3}\rho^2 + \tfrac{4}{33}\rho^4 - \tfrac{8}{1287}\rho^6)$$

Note that degeneracy also occurs in Eq. (8.54b); that is, for a given energy, different even l states or odd l states are possible for differing n. Thus the first few energy levels of the spherical harmonic oscillator may have the following number of independent eigenfunctions: $(n = 0, l = 0)$, only one independent wave function; $(n = 1, l = 1)$, $2l + 1 = 3$ different values of m are possible and therefore three independent wave functions; in the third level $(n = 1, l = 0)$ and $(n = 0, l = 2)$, a state of $1 + 5 = 6$ or sixfold degeneracy; and for $2n + l = 3$ a tenfold degeneracy. If $2n + l = k$, then the total degeneracy is given by

$$\sum_{m=0}^{(2k+1)/4} (2k + 1 - 4n)$$

Thus, the total degeneracy is given by

$$(k + 1)\left(\frac{k}{2} + 1\right) \sim \left(k + \frac{3}{2}\right)^2 \Big/ 2$$

Including the fact that two particles of spin $+\tfrac{1}{2}$ and $-\tfrac{1}{2}$ may occupy states having the same spatial wave function, we would expect nuclei with 2, $2 + 6 = 8$, $8 + 12 = 20$, and $20 + 20 = 40$ protons or neutrons to be especially stable if the harmonic oscillator potential is a reasonably good approximation to the nuclear potential. Nuclei having 2, 8, 20, 28, 50, 82, and 126 nucleons of the same kind are indeed found to have especially high binding energy. The discrepancy for numbers greater than twenty, present for any reasonable assumed potential (harmonic oscillator, square well, or whatever), was a big stumbling block to the early development of the theory.

and was finally resolved by introducing an interaction between the spin and orbital angular momentum of a nucleon, which will be discussed in more detail in Chapter 9.

The harmonic oscillator in one dimension is also of great interest because of its application to electromagnetic radiation in a rectangular box and other problems. Pauling and Wilson have presented a very lucid derivation of the one-dimensional harmonic oscillator which closely parallels the treatment here for a radial harmonic oscillator.* The wave equation for a one-dimensional harmonic oscillator is given by Eq. (8.47) with $l = 0$, $r = x$, and $R(r) = x^{-1}\psi(x)$. The solution is identical with Eq. (8.49) where $F(x)$ is taken to be an ordinary Hermite polynomial instead of a spherical Hermite polynomial. The energy eigenvalues are

$$E = \hbar\omega(n + \tfrac{1}{2}) \tag{8.55}$$

The wave functions are

$$\psi_n(x) = \left[\left(\frac{m\omega}{\pi\hbar}\right)^{1/2} \frac{1}{2^n n!}\right]^{1/2} e^{-m\omega x^2/2\hbar} F_n(\sqrt{m\omega/\hbar}\, x)$$

where

$$F_n(\sqrt{m\omega/\hbar}\, x) = \sum_{i=0,1}^{n} a_i(\sqrt{m\omega/\hbar}\, x)^i$$

$$a_{i+2} = -\frac{E/\hbar\omega - 2i - 1}{(i + 1)(i + 2)} a_i$$

The summation starts with 0 if n is even, 1 if n is odd.

Problem 8.11 Find the average values or expectations for x and x^2 in a one-dimensional oscillator and r in a spherically symmetric oscillator.

Problem 8.12 What is the probability of finding a particle of energy E within a distance less than a from the origin in a spherically symmetric harmonic oscillator potential if the particle's orbital momentum is zero?

Problem 8.13 Express $f(x) = e^{-2x^2}$ as a sum of harmonic oscillator wave functions in one dimension using the first five terms. (See the hint given in parentheses in Problem 8.4.)

* Pauling, L., and E. B. Wilson, Chap. 3, Section 11 (cited in Bibliography of this chapter).

Problem 8.14   A linear oscillator of mass m and angular frequency ω has its state represented by

$$\psi = e^{-m\omega x^2/\hbar^2}$$

What are the relative probabilities that its energy will be measured as $\hbar\omega/2$, $3\hbar\omega/2$, and $5\hbar\omega/2$? (See Problem 8.4.)

Problem 8.15   A linear oscillator of mass m and angular frequency ω has its state represented by

$$\psi = e^{-m\omega x^2/\hbar^2}$$

What is the probability that its momentum is larger than $\sqrt{\hbar m \omega}$?

BIBLIOGRAPHY

Pauling, L., and E. B. Wilson, *Introduction to Quantum Mechanics*. New York: McGraw-Hill Book Co., 1935. An exceptionally lucid and detailed treatment of the angular and radial wave functions of the hydrogen atom is given. The book also contains excellent tables of the hydrogen wave functions.

Jahnke, E., F. Emde, and F. Lösch, *Tables of Higher Functions*, 6th Ed. New York: McGraw-Hill Book Co., 1960. This is a standard reference work of formulas, tables, and graphs of many of the functions most frequently encountered in physics.

Morse, P., and H. Feshbach, *Methods of Theoretical Physics*. New York: McGraw-Hill Book Co., 1953. See page 622.

Schiff, L., *Quantum Mechanics*, p. 77 (cited in Bibliography of Chapter 7).

PERTURBATION THEORY

9

IN THE PREVIOUS CHAPTER we solved the Schroedinger equation exactly for several simple potential energy functions. In general, for any arbitrary potential, it is not possible to do this, of course, and we wish to take up in this chapter approximation methods of particular utility for problems in which the potential energy function does not differ much (i.e., is only slightly *perturbed*) from that in a simple, previously solved problem. There are many cases that may be handled by perturbation theory, such as (i) the electrons in a helium atom whose wave functions would exactly correspond to the wave function of the electron in a helium ion but for the slight disturbance or perturbation caused by the added repulsive interaction of the two electrons; (ii) an anharmonic oscillator for which the potential energy might contain a small dependence on powers of r higher than the second; (iii) the superposition of an externally applied electric or magnetic field on the usual Coulomb interaction between a proton and electron in the hydrogen atom; or (iv) two nucleons having a slight mutual interaction in a field of force caused by the combined effect of all the other nucleons in the nucleus.

In developing the perturbation theory we will have occasion to express various wave functions in the following manner:

$$\psi = \sum_n C_n \, \varphi_n,$$

where the C's are constants and where the φ's form a complete set of orthogonal functions like the trigonometric functions discussed earlier in connection with Fourier series. The orthogonality property, it will be recalled, is

$$\int \varphi_n{}^* \, \varphi_m \, d\tau = 0, \text{ if } n \neq m,$$

where the integration is extended over all space, and the completeness property means that any arbitrary ψ can be put into this form (If one of the φ_n's, say φ_1, were missing, the others would not form a complete set, since φ_1 itself could not be expressed in terms of them; for if it were, we would have

$$\varphi_1 = \sum_{n=2}^{\infty} C_n \, \varphi_n,$$

and consequently

$$\int |\varphi_1|^2 \, d\tau = \sum_{n=2}^{\infty} C_n \int \varphi_1{}^*\varphi_n \, d\tau = 0,$$

because of the orthogonal property. But this is impossible, since φ_1 would then be identically zero.) Let us agree to multiply each φ by a suitable constant so as to make it satisfy the condition

$$\int |\varphi_n|^2 \, d\tau = 1,$$

which is called the *normalization* condition. Then, for any ψ, we may calculate the coefficients C_n as follows: multiply the equation for ψ given above by $\varphi_m{}^*$ and integrate over all space.

$$\int \varphi_m{}^* \psi \, d\tau = \sum_n C_n \int \varphi_m{}^* \, \varphi_n \, d\tau = C_m,$$

where use has been made of the orthogonality and normalization conditions. Thus, given the function ψ, we can easily calculate C_m by evaluating the integral.

Let A be any Hermitian operator, with eigenvalues a_n and eigenfunctions φ_n:

$$A\varphi_n = a_n\varphi_n$$

It can then be shown, using the Hermitian property, that any two eigenfunctions, φ_n and φ_m, are orthogonal, provided that the eigenvalues a_n and a_m are distinct. If, however, φ_n and φ_m are two distinct eigenfunctions belonging to a single eigenvalue (the eigenvalue is degenerate in this case), then they need not be orthogonal, but they can be made so by a procedure to be explained presently. Thus, for any Hermitian operator, one obtains a set of orthogonal eigenfunctions, and in every case of interest it turns out to be a complete set. This applies in particular to the Hamiltonian operator, in which case the operator equation above is the time-independent Schroedinger equation. Thus the various sets of spherical wave functions derived in Chapter 8 all provide examples of complete orthogonal sets. A complete orthogonal set which has been normalized is called a *complete orthonormal set*.

Expression of the solution to a given wave equation in terms of a sum of wave functions which are solutions to a different wave equation is similar to the superposition of plane waves in simple Fourier analysis. Any function may be represented by a sum of sinusoidal functions having specified amplitudes. A single complicated electromagnetic disturbance can be represented as a sum or *superposition* of many simple plane waves of differing frequency whose amplitudes are found by the procedures given in Chapter 2. Any attempt to measure the intensity of a single wave present in the summation representing the single complicated disturbance would yield a positive result with a relative intensity identical with the square of the absolute value of the computed amplitude. Similarly we may represent a given quantum mechanical state by a *superposition of states* of the complete orthonormal set of solutions to a wave equation. It is most convenient to choose solutions to a wave equation closely corresponding to the actual wave equation we wish to solve or discuss. It would involve an unnecessary amount of labor to attempt, for example, to represent solutions to the wave equation for the electron in the hydrogen atom as a superposition of solutions to the wave equation for a particle in a rectangular box.

Eigenfunctions of the wave equation having different energy eigen-values are necessarily orthogonal, as we will now show. Let ψ_i and ψ_j be solutions of

$$\nabla^2\psi_i + \frac{2m}{\hbar^2}(E_i - V)\psi_i = 0$$

$$\nabla^2\psi_j{}^* + \frac{2m}{\hbar^2}(E_j - V)\psi_j{}^* = 0$$

(9.1)

Multiplying the terms in the first equation by $\psi_j{}^*$ and the terms in the second by ψ_i on the right, and subtracting the second equation from the first, we obtain

$$\psi_j{}^*\nabla^2\psi_i - (\nabla^2\psi_j{}^*)\psi_i + \frac{2m}{\hbar^2}(E_i - E_j)\psi_j{}^*\psi_i = 0$$

(9.2)

Integrating over the particle coordinates after representing the system in Cartesian coordinates, we obtain

$$\int_{-\infty}^{\infty}\int_{-\infty}^{\infty}\int_{-\infty}^{\infty}\left\{\psi_j{}^*\left(\frac{\partial^2\psi_i}{\partial x^2} + \frac{\partial^2\psi_i}{\partial y^2} + \frac{\partial^2\psi_i}{\partial z^2}\right)\right.$$
$$\left. - \left(\frac{\partial^2\psi_j{}^*}{\partial x^2} + \frac{\partial^2\psi_j{}^*}{\partial y^2} + \frac{\partial^2\psi_j{}^*}{\partial z^2}\right)\psi_i\right\} dx\,dy\,dz$$
$$+ \frac{2m}{\hbar^2}(E_i - E_j)\int_{-\infty}^{\infty}\int_{-\infty}^{\infty}\int_{-\infty}^{\infty}\psi_j{}^*\psi_i\,dx\,dy\,dz = 0$$

(9.3)

Since

$$\int_{-\infty}^{\infty}\left\{\psi_j{}^*\frac{\partial^2\psi_i}{\partial x^2} - \frac{\partial^2\psi_j{}^*}{\partial x^2}\psi_i\right\} dx = \int_{-\infty}^{\infty}\frac{\partial}{\partial x}\left\{\psi_j{}^*\frac{\partial\psi_i}{\partial x} - \frac{\partial\psi_j{}^*}{\partial x}\psi_i\right\} dx$$

$$= \left[\psi_j{}^*\frac{\partial\psi_i}{\partial x} - \frac{\partial\psi_j{}^*}{\partial x}\psi_i\right]_{-\infty}^{\infty} = 0$$

(9.4)

because of the boundary conditions on ψ, Eq. (9.3) reduces to

$$\frac{2m}{\hbar^2}(E_i - E_j)\int_{-\infty}^{\infty}\int_{-\infty}^{\infty}\int_{-\infty}^{\infty}\psi_j{}^*\psi_i\,dx\,dy\,dz = 0$$

(9.5)

Therefore the wave functions are orthogonal, for either $i = j$ and

$$\iiint \psi_j^* \psi_i \, d\tau = 1$$

or $i \neq j$ in which case $E_i \neq E_j$ and

$$\iiint \psi_j^* \psi_i \, d\tau = 0$$

The important case for which more than one state exists having the same energy eigenvalue will be treated later in Section 9.4.

9.1 FIRST-ORDER PERTURBATION THEORY

The Schroedinger wave equation

$$\nabla^2 \psi + \frac{2m}{\hbar^2}(E - V)\psi = 0 \tag{9.6}$$

may contain a potential energy term V which is only slightly different from the potential energy V^0 of a problem already solved. Since we should expect the solutions of Eq. (9.6) in a perturbation problem to differ very little from the solution found with V^0, we will use the wave functions for V^0 in attempting to expand the solutions to Eq. (9.6) in terms of known functions. For a given wave function ψ_i, let

$$\begin{aligned}
V &= V^0 + V' \\
E_i &= E_i^0 + E_i' \\
\psi_i &= \psi_i^0 + \psi_i' \\
H^0 &= -\frac{\hbar^2}{2m}\nabla^2 + V^0
\end{aligned} \tag{9.7}$$

where ψ_i^0 and E_i^0 are the solutions (eigenfunctions) and permitted energy values (eigenvalues) of the *unperturbed* Schroedinger equation

$$\nabla^2 \psi_i^0 + \frac{2m}{\hbar^2}(E_i^0 - V^0)\psi_i^0 = 0 \tag{9.8}$$

which may be written more compactly as

$$H^0\psi_i{}^0 = E_i{}^0\psi_i{}^0$$

The subscript i, distinguishing the i independent energy eigenvalues for the system, takes on different values with any change in the separate eigenvalues (e.g., n, l, m) of the wave function. (We are assuming here that there is only one eigenfunction corresponding to each energy eigenvalue and hence we need only one subscript; we will remove this restriction in a later section on degenerate levels.) The subscript i is necessary, as we have seen in Chapter 8, since each allowed value of E, E_i, results in a different ψ (for example, in a Hermite polynomial of different order, in the case of the harmonic oscillator). Substituting Eqs. (9.7) into Eq. (9.6) and grouping the result in ascending order of approximation, we obtain

$$[H^0 - E_i{}^0]\psi_i{}^0 + [(H^0 - E_i{}^0)\psi_i{}' + (V' - E_i{}')\psi_i{}^0]$$
$$+ \text{ second-order terms such as } (V_i{}' - E_i{}')\psi_i{}'] = 0 \tag{9.9}$$

The first bracket of terms is zero by Eq. (9.8), leaving us with only the second bracket of terms for a first-order approximation:

$$(H^0 - E_i{}^0)\psi_i{}' + (V' - E_i{}')\psi_i{}^0 = 0 \tag{9.10}$$

We now expand $\psi_i{}'$ in terms of the complete orthonormal set of solutions to Eq. (9.8), $\psi_j{}^0$. That is, we let

$$\psi_i{}' = \sum_{j=0}^{\infty} a_{ij}\psi_j{}^0 \tag{9.11}$$

which, with (9.10), gives

$$\sum_{j=0}^{\infty} (H^0 - E_i{}^0)a_{ij}\psi_j{}^0 + (V' - E_i{}')\psi_i{}^0 = 0 \tag{9.12a}$$

Note that from the second of Eqs. (9.8), Eq. (9.12a) may be rewritten

$$\sum_{j=0}^{\infty} (E_j{}^0 - E_i{}^0)a_{ij}\psi_j{}^0 + (V' - E_i{}')\psi_i{}^0 = 0 \tag{9.12b}$$

Multiplying Eq. (9.12b) by ψ_k^{0*} on the left and integrating over all space, we have

$$\sum_{j=0}^{\infty} (E_j^0 - E_i^0)a_{ij} \int \psi_k^{0*}\psi_j^0 \, d\tau + \int \psi_k^{0*} V'\psi_i^0 \, d\tau \tag{9.13}$$

$$-E_i' \int \psi_k^{0*}\psi_i^0 \, d\tau = 0$$

Since the ψ_i^0's are orthonormal, all but one term in the summation is zero, and Eq. (9.13) becomes

$$(E_k^0 - E_i^0)a_{ik} + \int \psi_k^{0*} V'\psi_i^0 \, d\tau - E_i' \int \psi_k^{0*}\psi_i^0 \, d\tau = 0 \tag{9.14}$$

If $k = i$, $E_i^0 - E_k^0 = 0$ and the shift in energy, E_i', due to the perturbation is

$$E_i' = \int \psi_i^{0*} V'\psi_i^0 \, d\tau \tag{9.15}$$

If $k \neq i$, the third term is zero and the a_{ik} are given by

$$a_{ik} = \frac{\int \psi_k^{0*} V'\psi_i^0 \, d\tau}{E_i^0 - E_k^0} \tag{9.16}$$

a_{ii} is not given by Eq. (9.16) but we already know its value; $a_{ii} = 1$ approximately since $\psi_i \sim \psi_i^0$.

$$a_{ii}^2 = 1 - \sum_{j \neq i} a_{ij}^2 = 1$$

up to and including terms of first order in the perturbation.

   Note that the first-order shift in energy from the unperturbed energy level, given in Eq. (9.15), is just the perturbed potential energy averaged over the unperturbed wave functions. We prefer to use a notation for the integrals in Eqs. (9.15, 9.16) in keeping with that of most authors, to wit,

$$V_{ki}' \equiv \int \psi_k^{0*} V'\psi_i^0 \, d\tau \tag{9.17}$$

obtaining the following expressions for the energy and wave functions of the perturbed system to first order:

$$E_i = E_i{}^0 + V_{ii}{}' \tag{9.18}$$

$$\psi_i = \psi_i{}^0 + \sum_{\substack{k=0 \\ k \neq i}}^{\infty} \frac{V_{ki}{}'}{E_i{}^0 - E_k{}^0} \psi_k{}^0 \tag{9.19}$$

Note that the effect of the perturbing potential is to give the particle a small but finite probability of occupying states of the unperturbed system other than the single unperturbed state it would be in if no perturbing potential existed.

Problem 9.1 Calculate the first-order energy correction and wave functions for the ground state of a particle in a one-dimensional square well with $V = -V_0 - \Delta$, $-a < x < 0$; $V = -V_0 + \Delta, 0 < x < a$; $V = 0$, $x < -a$ and $x > a$. ($\Delta < < V_0 \sim \infty$.)

Problem 9.2 Calculate the first-order energy correction and wave functions for the two lowest states of a particle in a perturbed one-dimensional square well with $V = -V_0 + cx$, $-a < x < a$. ($ca < < V_0 \sim \infty$.)

Problem 9.3 Calculate the first-order energy correction and wave functions for the two lowest states of a perturbed one-dimensional harmonic oscillator with $V = kx^2 + cx^3$. ($cx_1 < < k$ where x_1 is the classically permitted range of x for an oscillator with energy E.)

9.2 THE HELIUM ATOM

The total potential energy for the helium atom is

$$V = -\frac{Ze^2}{r_1} - \frac{Ze^2}{r_2} + \frac{e^2}{r_{12}} \tag{9.20}$$

where Z is 2 for the charge on the helium nucleus, r_1 and r_2 are the electron-nucleus separation distances for each electron, and r_{12} is the separation of the two electrons from each other. If derivatives with respect to a particular electron's position are denoted by a subscript 1 or 2, we can write the Schroedinger equation as

$$\nabla_1^2 \psi_i + \nabla_2^2 \psi_i + \frac{2m}{\hbar^2}(E_i - V_1^0 - V_2^0 - V')\psi_i = 0 \qquad (9.21)$$

where V_1^0 and V_2^0 are the hydrogen-like potential functions of the two electrons in the nuclear Coulomb field and their interaction energy is treated as a perturbation. Letting a superscript zero represent the zero-order wave function, the total wave function ψ_i^0 may be written as a product, $\psi_{1i}^0 \psi_{2j}^0$, of the wave functions of the two electrons taken separately, where ψ_{1i}^0 for example is a solution of

$$\nabla_1^2 \psi_{1i}^0 + \frac{2m}{\hbar^2}(E_{1i}^0 - V_1^0)\psi_{1i}^0 = 0 \qquad (9.22)$$

or

$$H_1^0 \psi_{1i}^0 = E_{1i}^0 \psi_{1i}^0$$

and $E_i^0 = E_{1i}^0 + E_{2i}^0$. The ψ_i^0 are the wave functions given in Chapter 8 in terms of product functions of Laguerre functions with associated spherical harmonics. The first approximation to the energy of the helium atom is given by substituting these wave functions into Eq. (9.18), whence

$$E_i = -\frac{mZ^2e^4}{2\hbar^2}\left(\frac{1}{n_1} + \frac{1}{n_2}\right) + \int \psi_{1n_1,l_1}^{0*} \psi_{2n_2,l_2}^{0*} \frac{e^2}{r_{12}} \psi_{1n_1,l_1}^{0} \psi_{2n_2,l_2}^{0} \, d\tau \qquad (9.23)$$

where n_1 and n_2 are the total quantum numbers of the unperturbed hydrogen-like wave functions. For the particular case of the ground state of the helium atom, $n_1 = n_2 = 1$ and $l = m = 0$. From Chapter 8, each of the one-electron wave functions is an exponential times the zero-order Laguerre polynomial (a constant), so that the normalized wave functions are

$$\psi_{11,0}^0 \psi_{21,0}^0 = \sqrt{\frac{Z^3}{\pi a_0^3}} \sqrt{\frac{Z^3}{\pi a_0^3}} \, e^{-\rho_1/2} \, e^{-\rho_2/2}$$

where $a_0 = \hbar^2/me^2$, $\rho_1 = 2Zr_1/a_0$, and the energy, E_1, is given by one half the first term in Eq. (9.23). Since in spherical coordinates, the volume element is

$$d\tau = r_1{}^2 \sin \theta_1 \, dr_1 \, d\theta_1 \, d\varphi_1 \, r_2{}^2 \sin \theta_2 \, dr_2 \, d\theta_2 \, d\varphi_2$$

the integral in Eq. (9.23) becomes

$$E_1' = \frac{Ze^2}{2(4\pi)^2 a_0} \int_0^\infty \int_0^\pi \int_0^{2\pi} \int_0^\infty \int_0^\pi \int_0^{2\pi} \left(\frac{e^{-\rho_1} e^{-\rho_2}}{\rho_{12}} \right)$$

$$\rho_1{}^2 \sin \theta_1 \, d\rho_1 \, d\theta_1 \, d\varphi_1 \, \rho_2{}^2 \sin \theta_2 \, d\rho_2 \, d\theta_2 \, d\varphi_2 \tag{9.24}$$

where $\rho_{12} = 2Zr_{12}/a_0$. Note that the integration is taken over a six-dimensional space, since the over-all wave function is concerned with the positions of two particles.

The integrand in Eq. (9.24) is essentially the electrostatic interaction energy between two shells of charge density $e^{-\rho_1}$ and $e^{-\rho_2}$. Its evaluation may be found in Appendix V of Pauling and Wilson (see Bibliography), but we will use the result so often that it is worthwhile to repeat the evaluation here.

To begin with let us consider just the integral over ρ_1. Let us consider further the potential at a point ρ due to a shell of thickness $d\rho$ at ρ_1; it is given by

$$\rho < \rho_1: \quad 4\pi\rho_1{}^2 \, \frac{e^{-\rho_1} \, d\rho_1}{\rho_1} = 4\pi\rho_1 e^{-\rho_1} \, d\rho_1$$

$$\rho > \rho_1: \quad 4\pi\rho_1{}^2 \, \frac{e^{-\rho_1} \, d\rho_1}{\rho} = \frac{4\pi}{\rho} \rho_1{}^2 e^{-\rho_1} \, d\rho_1$$

The total potential at ρ due to the infinite sphere of charge density $e^{-\rho_1}$ is given by

$$\phi(\rho) = 4\pi \int_0^\rho \rho_1 e^{-\rho_1} \, d\rho_1 + \frac{4\pi}{\rho} \int_\rho^\infty e^{-\rho_1} \rho_1{}^2 \, d\rho_1$$

$$= \frac{4\pi}{\rho} [2 - e^{-\rho}(\rho + 2)]$$

This is the potential at a point ρ due to the entire distribution of charge density $e^{-\rho_1}$. We can now evaluate the potential energy of interaction between the distributions of charge density $e^{-\rho_1}$ and $e^{-\rho_2}$. The potential energy per unit volume of a charge density $e^{-\rho_2}$ at a point ρ_2 is given by

$$\frac{4\pi}{\rho_2}[2 - (\rho_2 + 2)e^{-\rho_2}]e^{-\rho_2}$$

Hence, the second shell of charge $4\pi\rho_2^2 e^{-\rho_2} d\rho_2$ has a potential energy of

$$[(4\pi)^2\rho_2^2 e^{-\rho_2} d\rho_2/\rho_2][2 - e^{-\rho_2}(\rho_2 + 2)]$$

The total electrostatic potential of the sphere of charge density $e^{-\rho_2}$ due to the sphere with charge density $e^{-\rho_1}$ is the integral over ρ_2 of this expression:

$$(4\pi)^2 \int_0^\infty \rho_2 e^{-\rho_2}[2 - e^{-\rho_2}(\rho_2 + 2)] \, d\rho_2 = \tfrac{5}{4}(4\pi)^2$$

Therefore Eq. (9.24) yields a result of

$$E_1' = \frac{5}{4}\frac{Ze^2}{2a_0} = \frac{5}{4}\frac{mZe^4}{2\hbar^2}$$

and from Eq. (9.23) with $n_1 = n_2 = 1$,

$$E_1 = -\frac{me^4}{2\hbar^2}(2Z^2 - \tfrac{5}{4}Z) \qquad (9.25)$$

where $me^4/2\hbar^2$ is the ground state energy of the hydrogen atom and $2Z^2$ times this is the unperturbed ground state energy of the two helium electrons.

The correction to the zero-order ground state energy of the helium atom is, as we should suspect, large, being $\tfrac{5}{4}Z/2Z^2 = 5/8Z \sim 30\%$ of the zero-order energy, and is subtractive since the effect of the electron–electron interaction is to detract from or put up a shield reducing the electron–nucleus interaction. The approximation is found surprisingly good, however, in view of the size of the electron–electron interaction relative to the electron–nucleus interaction: the observed ground state energy of the helium atom is 78.6 volts, the zero-order calculated energy is 108.2 volts, and the first-order calculated energy is 74.4 volts. Thus, the zero-order calculation is found to be 38% too high while the first-order calculation is within 4.2 volts of the observed value, only 5% too low. In a later section, using the same wave functions we will take up a slightly different method which will reduce the error to only 2% of the observed value. The method yields better and better results for two-electron ions of heavier atoms as the electron–electron interaction becomes less important compared with the electron–nucleus interaction. Although the error is roughly the same (about -4 volts) for Li^+, Be^{++}, and B^{3+}, the ion binding energy increases rapidly with Z (it is proportional to Z^2) so that the first-order calculation of C^{4+} is only 0.4% too low.

Equation (9.23) is misleading in that it implies that for n_1 or $n_2 > 1$ each energy eigenvalue is associated with a unique eigenfunction specified by a pair of quantum numbers n, the total quantum number, and l, the orbital quantum number. Actually there are n^2 unique hydrogen-like wave functions associated with each energy level specified by total quantum number n. (There are $n - 1$ states of different l, and for each l, $2l + 1$ states of different m, all having the same energy.) Hence these n^2 states are mutually degenerate for hydrogen-like atoms. The energy level is a function of the sum of the radial and orbital quantum numbers and the simple theory breaks down when the energies of two different states with different quantum numbers are equal. Labeling states of different quantum numbers by different letters i and k in Eq. (9.19), we find that the energy denominator is zero for a degenerate level, giving infinite contributions of other states which have been assumed small in order for the approximation to be valid.

Problem 9.4 Calculate the first-order energy correction to the perturbed three-dimensional harmonic oscillator with potential energy

$$V = kr^2 + cr^3 + dr^4$$

Problem 9.5 Calculate the first-order energy correction and wave function for the ground state of the valence electron in the sodium atom if the unperturbed wave function is taken to be a hydrogen wave function with $n = 3$ and $l = m = 0$, and the atomic "core" of ten electrons is treated as a perfect screen resulting from the distribution of the electrons on a spherical shell of radius 0.2Å. $V = -10e/r$ for $r \leqslant 0.2$ Å, and beyond $r = 0.2$ Å the effective nuclear charge is $+e$. (Note that what you will obtain is an incomplete solution to this problem. See Problem 9.7.)

9.3 SECOND-ORDER PERTURBATION THEORY

When the first-order terms in Eq. (9.4) are zero or in general if a better approximation is needed, the second-order terms must be kept. We generalize the result further by assuming an additional correction to the potential, V'', so that our result may be applied also to nondegenerate cases in which a further, smaller physical effect may be estimated in addition to a larger perturbing potential. For this case we let

$$V = V^0 + V' + V''$$

$$E_i = E_i{}^0 + E_i' + E_i'' \qquad (9.26)$$

$$\psi_i = \psi_i{}^0 + \psi_i' + \psi_i''$$

and substitute into Eq. (9.1) to obtain for the second-order bracket in Eq. (9.9)

$$(H^0 - E_i{}^0)\psi_i'' + (V' - E_i')\psi_i' + (V'' - E_i'')\psi_i{}^0 = 0 \qquad (9.27)$$

Upon letting

$$\psi_i'' = \sum_l a_{il}'' \psi_l{}^0$$

and

$$\psi_i' = \sum_m a_{im}' \psi_m{}^0$$

multiplying Eq. (9.29) by $\psi_k{}^{0*}$ on the left, and integrating over all space, we obtain

$$(E_k{}^0 - E_i{}^0)a_{ik}'' + \sum_m a_{im}'(V_{km}' - E_i'\delta_{ki}) + (V_{ki}'' - E_i''\delta_{ik}) = 0 \qquad (9.28)$$

since

$$\int \psi_k{}^{0*}H^0\psi_l{}^0 \, d\tau = E_l \int \psi_k{}^{0*}\psi_l{}^0 \, d\tau = E_l\delta_{lk}$$

where it will be recalled that $\delta_{lk} = 0$ unless $l = k$ in which case $\delta_{lk} = 1$.

If $i = k$, since $V_{ii} = E'_i$ there results

$$E''_i = V''_{ii} + \sum_{\substack{m \neq i \\ m=0}}^{\infty} a'_{im} V'_{im}$$

$$= V''_{ii} + \sum_{\substack{m \neq i \\ m=0}}^{\infty} \frac{V'_{im} V'_{mi}}{E_i^0 - E_m^0} \tag{9.29}$$

When $i \neq k$, Eq. (9.28) yields the result that

$$a''_k = \frac{V''_{ki}}{E_i^0 - E_k^0} + \sum_{m \neq k}^{\infty} \frac{V'_{km} V'_{mk}}{(E_i^0 - E_k^0)(E_k^0 - E_m^0)} \tag{9.30}$$

Note that the second-order perturbation to the unperturbed energy consists of the second-order potential to the first order plus the first-order potential to the second order. The second-order potential may be some additional perturbation small compared to the first-order perturbation potential. The second term must be used whenever it happens that the first-order potential averaged over the unperturbed system gives zero. The presence of neighboring unperturbed levels, in which the particle can spend a small fraction of its time, contributes to the particle energy in this way, through the second term. The closer together the neighboring states lie in energy, the smaller the denominator becomes, and hence the greater their contribution becomes. We will return to second-order perturbation theory in Chapter 11.

Problem 9.6   A rigid body rotating in a plane has a wave function which satisfies the zero-order Schroedinger equation

$$\frac{1}{R^2} \frac{d^2 \psi_0}{d\varphi^2} + \frac{2m}{\hbar^2} E \psi_0 = 0$$

If the body has an electric dipole moment μ, its potential energy in an electric field F is given by $V = -\mu F \cos \varphi$. Calculate the ground state energy of this rotator in the applied field F, to second order.

Problem 9.7 If an electric field F parallel to the polar axis is applied to a hydrogen atom, the perturbing potential on the electron is

$$V' = -eF \cos \theta$$

Calculate the ground state energy to second order.

9.4 DEGENERATE LEVELS

Since degenerate levels occur with linearly independent functions, we first consider what the term linear independence means. Consider a set of functions $\psi_1(x)$, $\psi_2(x)$, ... $\psi_n(x)$, and form the linear combination

$$C_1 \psi_1(x) + C_2\psi_2(x) + \ldots + C_n \psi_n (x),$$

where the C's are constants not all equal to zero. If the C's can be chosen so that this combination vanishes identically, then the set of functions is said to be linearly dependent. Otherwise it is linearly independent.

Independent eigenfunctions (e.g. with different l and m) having the same energy eigenvalue are said to belong to a degenerate energy level. We have previously mentioned the difficulties they present to perturbation theory, necessitating a modification of the theory. The modification is straight-forward and elementary but requires somewhat formidable notation. Although degenerate eigenfunctions are not necessarily orthogonal, in contrast to eigenfunctions with differing energy eigenvalues, an orthogonal set of degenerate eigenfunctions may be obtained by taking appropriate sums of the linearly independent eigenfunctions belonging to the same energy level. Members of the orthogonal set χ_1, χ_2 must fulfill such conditions as the following, for twofold degenerate systems:

$$\chi_1 = b_{11}\psi_1 + b_{12}\psi_2 \qquad \chi_2 = b_{21}\psi_1 + b_{22}\psi_2$$

that is, the χ_i's are linear combinations of the ψ_i's, which are caused to be orthogonal by requiring that

$$\int \chi_1{}^*\chi_2 \, d\tau = b_{11}{}^*b_{21}\int \psi_1{}^*\psi_1 \, d\tau + b_{11}{}^*b_{22}\int \psi_1{}^*\psi_2 \, d\tau$$

$$+ b_{12}{}^*b_{21}\int \psi_2{}^*\psi_1 \, d\tau + b_{12}{}^*b_{22}\int \psi_2{}^*\psi_2 \, d\tau = 0$$

that is, the χ_i's are orthogonal even though the ψ_i's may not be orthogonal. In general, where K is the order of the degeneracy,

$$\chi_i = \sum_{k=1}^{K} b_{ik}\psi_k \tag{9.31}$$

and

$$\int \chi_i^*\chi_j \, d\tau = \sum_{m}^{K}\sum_{n}^{K} b_{im}{}^*b_{jn}\int \psi_m{}^*\psi_n \, d\tau = \delta_{ij} \tag{9.32}$$

This does not completely specify the b_{ik}; these are formed from additional conditions which we will develop.

We could continue as in the previous sections to label a unique state with only one subscript on the eigenfunctions and only two subscripts on the coefficients (the a_{ij}'s in Eq. (9.11)). However, it is more convenient to split the eigenfunctions into subgroups according to energy eigenvalues. Two subscripts will therefore be needed to specify a state: the first subscript to label a particular subgroup of wave functions having the same energy eigenvalue, the second subscript to label a particular wave function in this subgroup. This labeling scheme means that the coefficients (the a_{ij}'s) must now have four subscripts: since two subscripts are now needed to specify a state, four subscripts will be needed to relate two states to each other. Thus $a_{jk,ni}$, for example, will mean the contribution of the ith basic wave function having the energy eigenvalue E_n to the kth state having the energy eigenvalue E_j (see Eq. 9.11a following). Equations (9.8) and (9.10) become, for degenerate energy levels,

$$H^0\chi_{ik}{}^0 = E_i{}^0\chi_{ik}{}^0 \tag{9.8a}$$

$$(H^0 - E_i{}^0)\chi_{ik}' + (V' - E_i')\chi_{ik}{}^0 = 0 \tag{9.10a}$$

Initially our unperturbed system may be represented in terms of wave functions which are not orthogonal $\psi_{i'k'}{}^0$'s but from which orthogonal wave functions may be constructed by the use of Eqs. (9.31) and (9.32). In terms of the original wave functions the χ'_{ik} may be written

$$\chi'_{ik} = \sum_{i',n} a_{ik,i'n}\chi_{i'n}{}^0 = \left(\sum_{i',n} a_{ik,i'n} \sum_{k'} b_{i'n,i'k'} \right)\psi_{i'k'}{}^0$$

$$= \sum_{i',n,k'} a_{ik,i'n}b_{i'n,i'k'}\psi_{i'k'}{}^0 = \sum_{i',k'} a_{ik,i'k'}\psi_{i'k'}{}^0 \qquad (9.11a)$$

where

$$a_{ik,i'k'} \equiv \sum_{n} a_{ik,i'n}b_{i'n,i'k'}$$

The subscripts i' on the $b_{i'n,i'k'}$ will be dropped to simplify the notation, with the understanding that the $b_{n,k'}$ are for one of the particular wave functions $\chi_{i'n}$ having the energy $E_{i'}$. Since

$$\chi_{ik}{}^0 = \sum_{k'} b_{kk'}\psi_{ik'}{}^0$$

similar to Eq. (9.12b) we now have

$$\sum_{i',k'} a_{ik,i'k'}(E_{i'}{}^0 - E_i{}^0)\psi_{i'k'}{}^0 = \sum_{k''} b_{kk''}(E_i' - V')\psi_{ik''}{}^0 \qquad (9.12c)$$

(Note that the dummy index on the right-hand side is labeled with a double prime to distinguish it from k'.) Multiplying Eq. (9.12c) by $\psi_{ij}{}^{0*}$ on the left and integrating over all coordinate space we obtain

$$\sum_{i',k'} a_{ik,i'k'}(E_{i'}{}^0 - E_i{}^0)\int \psi_{ij}^{0*}\psi_{i'k'}^0 \, d\tau = \sum_{k''} b_{kk''}(E_i'I_{jk''} - V'_{jk''}) \qquad (9.14a)$$

where

$$I_{jk''} = \int \psi_{ij}^{0*}\psi_{ik''}{}^0 \, d\tau$$

and

$$V'_{jk''} = \int \psi_{ij}^{0*} V'\psi_{ik''}{}^0 \, d\tau \qquad (9.15a)$$

The left-hand side of Eq. (9.14a) is zero because of the orthogonality of the ψ^0's for $i \neq i'$ and because the coefficient $E_{i'}{}^0 - E_i{}^0$ is zero if $i = i'$. Thus we arrive at a set of simultaneous equations in the unknown $b_{kk''}$, there being as many unknown $b_{kk''}$ as the number of the degeneracy (number of degenerate states) of the ith energy level:

$$\sum_{k''} b_{kk''}(V'_{jk''} - E_i' I_{jk''}) = 0 \qquad (9.33)$$

Unless the determinant of the coefficients of the $b_{kk''}$ in Eq. (9.33) is zero, however, all the $b_{kk''}$ must be identically zero in order for Eqs. (9.33) to be satisfied. If we had started with orthogonal wave functions we would have $I_{jk''} = \delta_{jk''}$ and instead of Eqs. (9.33)

$$\sum_{k''} b_{kk''}(V'_{jk''} - E_i' \delta_{ik''}) = 0 \qquad (9.34)$$

whose determinant for the coefficients is

$$\begin{vmatrix} (V_{11}' - E_i') & V_{12}' & \cdots & V_{1n}' \\ V_{21}' & (V_{22}' - E_i') & \cdots & V_{2n}' \\ \cdot & \cdot & & \cdot \\ \cdot & \cdot & & \cdot \\ \cdot & \cdot & \cdots & (V_{nn}' - E_i') \end{vmatrix} = 0 \qquad (9.35)$$

the degeneracy in the ith energy level being n-fold. It frequently happens that the perturbation potential taken between two states is zero unless the two states are the same. That is to say, $V'_{jm} = \delta_{jm} V_{jj}$. For example, this would occur, if the perturbing potential did not depend on angle, the original wave functions being for a spherically symmetric potential. In this case the orthogonality of the Y_l^m would result in $V'_{jm} = \delta_{jm} V'_{jj}$. Equation (9.35) then becomes diagonal,

$$\begin{vmatrix} (V_{11}' - E_i') & 0 & \cdots & 0 \\ 0 & (V_{22}' - E_i') & \cdots & 0 \\ \cdot & \cdot & & \cdot \\ \cdot & \cdot & & \cdot \\ \cdot & \cdot & \cdots & (V_{nn}' - E_i') \end{vmatrix} = 0 \qquad (9.35a)$$

with the particularly simple form

$$(V_{11}' - E_i')(V_{22}' - E_i') \cdots (V_{nn}' - E_i') = 0$$

whose solutions are

$$E_i' = V_{11}', V_{22}', \ldots, V_{nn}' \qquad (9.36)$$

for a level of n-fold degeneracy.

The original wave functions for this case are called the correct zero-order wave functions. Clearly, the problem is immensely simplified if we can find at the start a representation in which the matrix Eq. (9.34) is diagonal, that is, a representation in which all the matrix elements not on the diagonal are zero. A spherically symmetric perturbing potential for a particle in a rectangular box, for example, would result in a nondiagonal matrix, if the original wave functions chosen were for the particle in a rectangular box. Hence, these wave functions would be an inappropriate choice for this problem.

Note that Eq. (9.36) with the V_{jk}' being given by Eq. (9.15a) is essentially the same result as that obtained with nondegenerate perturbation theory (Eq. 9.15) with the difference that in degenerate perturbation theory the perturbation potential energy V' must be evaluated for not just one eigenfunction belonging to E_i^0 but between all eigenfunctions in the same energy level E_i^0.

Problem 9.8 The $n = 3$ state for the hydrogen atom is actually a 9-fold degenerate state. Without bothering to solve the integrals obtained along the diagonal, set up the perturbation determinant for Problem 9.5. What are the off-diagonal elements?

9.5 FINE STRUCTURE

A spinning electron in an atom, which we will discuss in more detail in Chapter 12, has a magnetic dipole moment due to its charge and intrinsic rotation. If the electron's motion about the nucleus is rotational (i.e., $l \neq 0$), then another magnetic moment arises because such motion is in effect another current loop. The interaction of these two magnetic dipoles gives rise to the so called spin–orbit force or potential which may be calculated by means of perturbation theory. The spin-orbit potential is shown in Section 12.2 to be

$$V' = \frac{1}{r} \frac{dV^0}{dr} \frac{\hbar^2 \boldsymbol{\sigma} \cdot \mathbf{1}}{4m^2c^2} \qquad (9.37)$$

where $\boldsymbol{\sigma}$ denotes twice the angular momentum operator of the electron spin in units of $\hbar$ and $\mathbf{1}$ denotes the electron orbital momentum operator in units of $\hbar$. This very weak interaction results in a minute shift in the atomic energy level, and in the case of atomic spectroscopy results in spectroscopic fine structure. We discuss atomic spin–orbit interaction here because of the role of the spin–orbit interaction in atomic and nuclear shell theory. To date the cause of the spin–orbit interaction in nuclear physics is not known; its magnitude in the nucleus is of the order of a million volts and it is responsible for the fact that the correct order of nuclear levels cannot be calculated on the basis of any simple short-range central potential such as a square well or harmonic oscillator potential.

The total angular momentum of an electron in the atom is a constant of the motion (that is, it does not change with time) and is the sum of the orbital angular momentum and the spin, $\mathbf{J} = \mathbf{1} + \frac{1}{2}\boldsymbol{\sigma}$, the electron spin $\frac{1}{2}\boldsymbol{\sigma}$ having one half unit of angular momentum in units of $\hbar$. The projection of the electron spin on any axis (including the orbital momentum axis) is quantized and is either $+\frac{1}{2}$ or $-\frac{1}{2}$, in units of $\hbar$. As we know, the eigenvalue of the square of any angular momentum operator is, for example, $\mathbf{1}^2 = l(l + 1)$, in units of $\hbar^2$. We have the identity

$$\mathbf{J}^2 = \mathbf{1}^2 + \boldsymbol{\sigma} \cdot \mathbf{1} + (\boldsymbol{\sigma}/2)^2$$

whence, since the operator $(\boldsymbol{\sigma}/2)^2$ operating on the electron spin wave function has the eigenvalue $\frac{1}{2}(\frac{1}{2} + 1)$,

$$\boldsymbol{\sigma} \cdot \mathbf{1} = \mathbf{J}(\mathbf{J} + 1) - l(l + 1) - \tfrac{3}{4} \qquad (9.38)$$

so that $\boldsymbol{\sigma} \cdot \mathbf{1} = +l$ for $J = l + \frac{1}{2}$ (i.e., the case in which electron spin and orbital momentum are in the same direction) and $\boldsymbol{\sigma} \cdot \mathbf{1} = -(l + 1)$ for $J = l - \frac{1}{2}$ (i.e., the spin and orbital momentum are opposed). If $l = 0$, an s state, $\mathbf{1} = 0$ and $\boldsymbol{\sigma} \cdot \mathbf{1} = 0$, not -1.

As an example let us calculate the fine structure of the $3d$ ($n = 3$, $l = 2$) state in the hydrogen atom. In this case V^0 is a simple Coulomb potential between electron and proton, and the radial part of ψ^0 is the associated Laguerre function of degree $n - l - 1 = 0$ and order $2l + 1 = 5$, described in Chapter 8. Since the angular part of ψ is normalized and the potential does not depend on the angles, the perturbed energy is calculated by simply integrating over r. The original wave functions are orthogonal and are the correct wave functions, as may be ascertained by observing that the off-diagonal matrix elements V_{jk}' give 0 between states of different l belonging to the same energy level, due to the orthogonality of states of different l and the angular independence of the perturbation potential. Hence $k = j$. The subscript j in this instance labels a $3d$ state.

$$V'_{jj} \equiv V'_{3d,3d} = \int_0^\infty \frac{4}{81\sqrt{30}} \left(\frac{me^2}{\hbar^2}\right)^{5/2} r^2 e^{-(me^2/3\hbar^2)r} \left[\frac{e^2\hbar^2}{4m^2c^2} \frac{\sigma \cdot 1}{r^3}\right]$$
$$\times \frac{4}{81\sqrt{30}} \left(\frac{me^2}{\hbar^2}\right)^{5/2} r^2 e^{-(me^2/3\hbar^2)r} r^2 \, dr \tag{9.39}$$

By integration we have

$$E' = V'_{3d,3d} = -\frac{1}{810} \frac{me^4}{2\hbar^2} \frac{e^4}{\hbar^2 c^2} \sigma \cdot 1 = -\frac{1}{810} W_H \alpha^2 \sigma \cdot 1$$

where W_H is the binding energy of the ground state of the hydrogen atom, $\alpha = e^2/\hbar c$ and is called the fine structure constant, and $\sigma \cdot 1 = +2$ or -3 depending on the orientation of the spin relative to the orbital momentum axis, so that the energy would be split into two levels having an energy difference of $(5/810) W_H \alpha^2$.

As we have already noted, all the states in hydrogen having the same total quantum number have the same energy eigenvalue (are mutually degenerate), since the energy is a function of the sum of the radial and orbital quantum numbers. These degenerate states are split up and separated by the magnetic interaction of the electron spin with the electron orbital motion; each separate value of l gives two new slightly shifted levels. In our example where $n = 3$, there would result five levels where before there was one level: two levels for the $3d$ state $l = 2$, two less separated levels for the $3p$ state $l = 1$, and one unshifted level for the $3s$ state $l = 0$. The degeneracy in the projection of l on a given axis, m, remains. Note that in our example we computed only one of the three integrals for the three-fold degeneracy $(3d, 3p, 3s)$ between states of different l but same total quantum number $(n = 3)$ for all the different values of $E_i{}'$ in Eq. (9.36). For atoms other than hydrogen, the central Coulombic potential is modified by the shielding effect of the inner or core electrons, and this effect is very much larger than the spin–orbit interaction. In the case of hydrogen-like alkali atoms, for example, the potential is nearly Coulombic

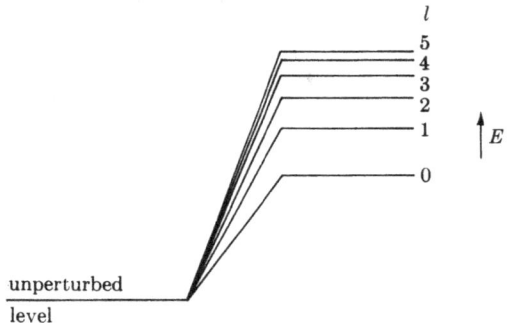

Figure 9.1 *Illustration of the way a single valence level of total degeneracy n^2 is broken up into n levels, each having a degeneracy $(2l + 1)$, by the effect of shielding by the core electrons in an alkali atom.*

for the outer electrons and their wave functions nearly hydrogenic. A level of total quantum number n, however, is split into n different levels corresponding to different l's. Since the probability of finding an electron at small r in a Coulomb potential decreases as l increases even though n, the total quantum number, remains the same (as illustrated in Figure 8.8); therefore outer electrons in alkali atoms represented by hydrogen wave functions of small l penetrate further inside the electron cloud shielding the nucleus than do outer electrons represented by states of larger l. Hence states of smaller l have lower energy than states of larger l having the same n, and the degeneracy is broken up as shown in Figure 9.1.

9.6 VARIATIONAL METHOD

Using the same form of the zero-order wave functions as used in perturbation theory we may obtain a better estimate of the ground state or lowest energy of a perturbed system with little additional labor, by means of the variational method. The method rests on noting that the integral for the expectation of the energy operator in terms of an arbitrary trial wave function, Φ, (which is, in general, not the lowest eigenfunction of the system) is given by

$$E = \int \Phi^* \left(\frac{-\hbar^2}{2m} \nabla^2 + V \right) \Phi \, d\tau = \int \Phi^* H \Phi \, d\tau \qquad (9.40)$$

and overestimates the energy of the lowest state of the system. We first proceed to prove this statement. Let Φ be expanded in terms of the complete orthonormal set comprised of the true wave functions of the system:

$$\Phi = \sum_i a_i \psi_i \qquad (9.41)$$

and similarly for Φ^*. Then

$$E = \sum_i \sum_j a_i^* a_j \int \psi_i^* H \psi_j \, d\tau \qquad (9.42)$$

Since $H\psi_j = E_j \psi_j$, and the ψ's are orthogonal, Eq. (9.42) becomes

$$E = \sum_j a_j^* a_j E_j \qquad (9.43)$$

The sum over all j of $a_j^* a_j$ is 1, and therefore $E \geq E_0$; E is equal to E_0 only if the a_j are zero for all $j \neq 0$, in which case $\Phi = \psi_0$ which means that we have chosen the true wave function for Φ. The worse the choice made for Φ, the greater will be the admixture of states above the ground state, leading to higher estimates of E.

In general, Φ should be chosen to resemble as closely as possible the form to be expected on physical grounds. For example, for nuclear problems a rapidly attenuating wave function should be used, such as an exponentially decaying wave function. A good approximate wave function is a known wave function appropriate to a very similar potential, that is, an unperturbed ground state wave function whose potential energy is assumed only slightly different from the actual potential energy. The only alteration required in the known wave function is to make Φ a function of some unspecified parameter, say α. The integral in Eq. (9.40) is then calculated as a function of α and the result minimized with respect to α, the resultant E being the

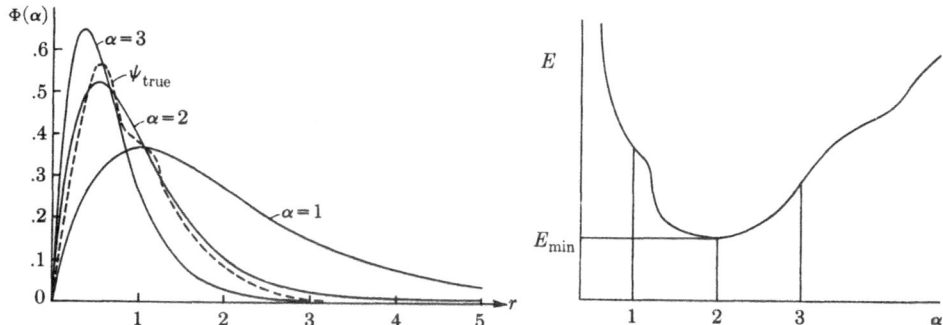

Figure 9.2 *Illustration of how an approximate wave function, $\Phi = re^{-\alpha r}$, may be adjusted to most closely represent an illustrative true wave function ψ_{true} (dashed line) by choice of the value of the parameter α that minimizes E given by Eq. (9.44), $\int \Phi H \Phi \, d\tau / \int \Phi^2 \, d\tau.$*

closest estimate obtainable with the chosen wave function. If we use unnormalized wave functions, Eq. (9.40) may be rewritten

$$E = \frac{\displaystyle\int \Phi^*(\alpha, \mathbf{r})\left(\frac{-\hbar^2}{2m}\nabla + V\right)\Phi(\alpha, \mathbf{r}) \, d\tau}{\displaystyle\int \Phi^*(\alpha, \mathbf{r})\Phi(\alpha, \mathbf{r}) \, d\tau} \tag{9.44}$$

For an illustration of what the variational method attempts to accomplish, see Figure 9.2 where trial wave functions $\Phi = re^{-\alpha r}$ are plotted for three different values of α. Φ represents an unknown wave function ψ_{true} which might have the form shown in the figure. The value of α for a best fit is found by minimizing the calculated energy given by $\int \Phi H \Phi \, d\tau / \int \Phi^2 \, d\tau$, as illustrated.

As a small example we attempt another calculation of the ground state of the helium atom, whose potential energy is given by Eq. (9.20). For Φ we choose the normalized zero-order wave functions used earlier in the chapter (e.g., in Eq. 9.24) but we let the nuclear charge be an undetermined parameter, Z', representing an effective nuclear charge which is less than the actual charge, owing to the screening of the attraction of any one electron to the nucleus by the presence of the other electron. Thus

$$\Phi(Z', \mathbf{r}_1, \mathbf{r}_2) = \Phi_1 \Phi_2 = \left(\frac{Z' m e^2}{\pi \hbar^2} \right)^3 e^{-(Z' m e^2 / \hbar^2)(r_1 + r_2)} \qquad (9.45)$$

where the subscripts 1, 2 refer to the two electrons. [Φ_1 and Φ_2 satisfy the hydrogen-like wave equation

$$\left(\frac{\hbar^2}{2m} \nabla_1^2 + \frac{Z' e^2}{r_1} \right) \Phi_1 = Z'^2 W_H \qquad (9.46)$$

where W_H has been defined in Eq. (9.39) as $m e^4 / 2 \hbar^2$. Note that

$$V^0 = - \frac{Z' e^2}{r_1} - \frac{Z' e^2}{r_2} - (Z - Z') \left(\frac{e^2}{r_1} + \frac{e^2}{r_2} \right) \Bigg] \qquad (9.47)$$

Substitution of Eqs. (9.45)–(9.47) into Eq. (9.44) yields

$$E = -2Z'^2 W_H + (Z' - Z) e^2 \int \Phi^* \left(\frac{1}{r_1} + \frac{1}{r_2} \right) \Phi \, d\tau$$

$$+ \int \Phi^* \frac{e^2}{r_{12}} \Phi \, d\tau \qquad (9.48)$$

The first integral is trivial, resulting in $(Z' - Z)[4Z' W_H]$. The second integral has already been found in obtaining Eq. (9.25) to be $\frac{5}{4} W_H Z'$; hence

$$E = \{ -2Z'^2 + 4Z'(Z' - Z) + \tfrac{5}{4} Z' \} W_H$$

Minimizing with respect to Z', that is, setting $\partial E / \partial Z' = 0$, we find that

$Z' = Z - \frac{5}{16}$ so that

$$E = -2(Z - \tfrac{5}{16})^2 W_H$$
$$= (-2Z^2 + \tfrac{5}{4}Z - \tfrac{50}{256}) W_H \qquad (9.49)$$
$$= -76.9 \text{ volts}$$

Our previous result, -74.4 volts, was given by the first two terms of Eq. (9.49). The extra term, $-\frac{50}{256} W_H = -2.5$ volts, is an appreciable fraction of the 4.2 volts difference from the observed result of -78.6 volts.

Problem 9.9   Find the ground state of the linear anharmonic oscillator whose potential energy is $V = kx^2 + 0.1k^2x^4$, using the trial wave function $\psi = ce^{-\alpha x^2}$.

Problem 9.10   Assume a proton to be not a point charge but a sphere with radius a of 10^{-13} cm; then find the ground state energy of the hydrogen atom if the positive charge is distributed in such a way as to cause (use ground state hydrogenic wave functions for trial functions) $V(r) \sim -e^2/(r + a)$.

Problem 9.11   Find the ground state energy of a particle in a spherically symmetric square well for a particle mass $M = 1.6 \cdot 10^{-24}$ g, $V_0 = -20$ Mev, and a well radius of $1.5 \cdot 10^{-13}$ cm. Use a trial wave function of $\psi = (c/r) e^{-\alpha r}$.

Problem 9.12   When tritium, H^3, beta-decays to He^3, the single atomic electron suddenly finds the positive nuclear charge doubled. If the initial tritium atom was in its ground state, what is the probability of finding the He$^+$ ion in its ground state instantaneously following the decay? (That is, take the ground state hydrogen wave function and express it as a series of helium ion wave functions (hydrogenic with $Z = 2$) and find the coefficient of the first term.)

BIBLIOGRAPHY

Pauling, L., and E. B. Wilson, cited in Chapter 8.

UP TO THIS POINT our attention has been directed to static systems undergoing no temporary interactions with other systems. Since the systems have not experienced outside forces and thus have exchanged no energy with other systems, the energy has remained constant. Under such circumstances involving time-invariant potentials, the complete wave equation (4.18),

$$\left(-\frac{\hbar^2}{2m}\nabla^2 + V \right)\Psi = H\Psi = -\frac{\hbar}{i}\frac{\partial\Psi}{\partial t} \tag{10.1}$$

is readily simplified, since the left-hand side does not involve the time, and therefore the space and time variables are separable. The time-dependent part of Ψ satisfies the differential equation

$$\frac{1}{\Psi(\mathbf{r},t)}\frac{\hbar}{i}\frac{\partial\Psi(\mathbf{r},t)}{\partial t} = -E \tag{10.2}$$

where E is a constant which was shown in Chapter 5 to be the energy of the system. The solution of Eq. (10.2) is

$$\Psi(\mathbf{r},t) = \psi(\mathbf{r})e^{-(t/\hbar)Et} \tag{10.3}$$

After substituting (10.3) in (10.1) and then dividing out the time-dependent part of Ψ, we are left with the space-dependent part of Ψ, to which we have devoted our attention in previous chapters:

$$H\psi = E\psi \tag{10.4}$$

There are many problems of paramount significance, however, in which an isolated static system is acted on by some external time-dependent disturbance. For example, a hydrogen atom in its ground state might experience an electromagnetic wave or photon incident upon it. The photon could cause a transition of the atom to an excited or ionized state. Calculation of the probability or cross-section of transition due to the absorption of a photon is of outstanding physical importance.

For such a problem, the space and time variables in Eq. (10.1) are not separable since H or V, which includes the effect of the incident photon, is time-dependent. The resulting differential equation is of such complexity that in the past it has been solved entirely by the use of time-dependent perturbation theory. This has been a truly desperate measure in many problems in which the perturbation is not small (e.g., nuclear interactions), and that it has been virtually the only tool available is an indication of the difficulty of time-dependent problems including static problems which have been treated as time dependent perturbations. A major difficulty in theoretical physics, particularly with respect to the problem of nuclear forces, has been the lack of a suitable alternative approximation technique when the time-dependent potential is too large for time-dependent perturbation approximations to converge rapidly or at all. The theory will be applied here only to electromagnetic interactions for which, fortunately, the approximations converge very rapidly. Electromagnetic interactions are very weak compared with nuclear interactions.

10.1 FIRST-ORDER TIME-DEPENDENT TRANSITION PROBABILITY

Dirac derived the time-dependent perturbation approximation technique at almost the same time that Schroedinger derived the time-independent theory (Chapter 9). As will be seen shortly, the two methods are very similar and have many parallels. In the time-dependent theory the system might be assumed initially in a time-independent state whose wave function satisfies the zero-order, time-independent wave Eq. (10.4); then the system is socked momentarily by an outside disturbance, after which it is left in a possibly different state with a new solution to Eq. (10.4). To phrase the problem more exactly: A static system whose wave function satisfies Eq. (10.4) suddenly experiences a perturbing potential (possibly representing an incident photon) for a short time, during which its wave function is an inseparable function of time and space; it is then restored to its previous static condition, except for a possibly changed eigenvalue (e.g., energy or angular momentum) and eigenfunction. Since the perturbing potential is assumed small compared with the initial energy of the system, the time-dependent wave functions are expanded in terms of the complete orthonormal set of static wave functions. The coefficients of the series must, of course, be time-dependent because of the effect of the perturbing potential.

The wave functions of the initial unperturbed static system satisfy the wave equation for the static potential,

$$H_0 \Psi_n = \left(\frac{-\hbar^2}{2m} \nabla^2 + V \right) \Psi_n = -\frac{\hbar}{i} \frac{\partial \Psi_n}{dt} = E_n \Psi_n \qquad (10.5)$$

For the duration of the perturbation, the appropriate complete wave equation is

$$(H_0 + H') \Psi = -\frac{\hbar}{i} \frac{\partial \Psi}{\partial t} \qquad (10.6)$$

where H' represents the small, time-dependent perturbation. For example H' could be the result of a relatively constant electric field due to the incidence of a photon having a frequency small compared with the electron's orbital frequency. We could take the case where H' does not contain the time explicitly. The perturbation begins at time zero and ends at some later time. The solution to Eq. (10.6) at any particular instant of time t may be written in terms of the complete orthonormal set of wave function solutions to Eq. (10.5):

$$\Psi(\mathbf{r}, t) = \sum_{n=0}^{\infty} a_n(t) \psi_n(\mathbf{r}) e^{-(t/\hbar)E_n t} \qquad (10.7)$$

where the $a_n(t)$ are functions only of the time and the $\psi_n(\mathbf{r})$ are functions only of the coordinates. If we were to measure some physical property of the system, e.g. an eigenvalue such as the energy, the probability of obtaining a particular value of the energy E_k of the state ψ_k would be given by $|a_k(t)|^2$. In other words the probability of finding the system in the state ψ_k at time t is given by $|a_k(t)|^2$. The solution to any problem is found, of course, by determining the time-dependent coefficients $a_n(t)$.

Just as in the time-independent perturbation theory, equations for the coefficients result from substituting Eq. (10.7) into Eq. (10.6):

$$\sum_n a_n(t) H_0 \psi_n\, e^{-(i/\hbar)E_n t} + \sum_n a_n(t) H' \psi_n\, e^{-(i/\hbar)E_n t}$$

$$= -\frac{\hbar}{i} \sum_n \dot{a}_n(t) \psi_n\, e^{-(i/\hbar)E_n t} + \sum_n a_n(t) \psi_n E_n\, e^{-(i/\hbar)E_n t} \qquad (10.8)$$

Utilizing Eq. (10.5) to eliminate the first and last terms, we obtain

$$i\hbar \sum_n \dot{a}_n(t) \psi_n\, e^{-(i/\hbar)E_n t} = \sum_n a_n(t) H' \psi_n\, e^{-(i/\hbar)E_n t} \qquad (10.9)$$

We next multiply Eq. (10.9) by $\Psi_k^*(\mathbf{r}, t)$ on the left and integrate over all configuration space. The orthogonality of the wave functions requires all resulting integrals on the left-hand side of Eq. (10.9) to be zero except for the kth term in the sum; hence

$$i\hbar \dot{a}_k(t) = \sum_n a_n(t) H'_{kn}\, e^{(i/\hbar)(E_n - E_k)t} \qquad (10.10)$$

where

$$H'_{kn} \equiv \int \psi_k^*(\mathbf{r}) H' \psi_n(\mathbf{r})\, d\tau \qquad (10.11)$$

Equation (10.10) is an infinite set of simultaneous equations for the $a_k(t)$.

If the perturbation is small and acts for times sufficiently short that the final wave function of the system has little probability of differing from that for the initial state, an approximate solution to Eq. (10.10) is readily found. Suppose the initial state of the system is ψ_{n_0}, that is, at $t = 0$ all the $a_n(0)$ are zero except a_{n_0} which is unity. In this case, Eq. (10.10) becomes

$$i\hbar \dot{a}_k(t) = H'_{kn_0} \exp\left\{-(i/\hbar)(E_{n_0} - E_k)t\right\} \qquad (10.12)$$

$$a_k(t) = \frac{1}{i\hbar} \int_0^t H'_{kn} \exp\left\{-\frac{i}{\hbar}(E_{n_0} - E_k)t\right\} dt \qquad (10.13)$$

If H'_{kn_0} is a nearly constant function of the time, that is, if it does not change appreciably over one cycle of the oscillation of the exponential function, the transition probability will be small, due to cancellations from times of opposing phase. That is, the exponential in (10.13) causes equal negative and positive contributions of the integrand to the integral if H'_{kn_0} is nearly constant, and the value of the integral will be nearly zero. If H'_{kn_0} is caused by electromagnetic radiation, $e^{i\omega t}$, or any other periodic motion, the transition probability will be large only for disturbance periods given by

$$T \text{ (max. disturbance)} = h/(E_{n_0} - E_k) \qquad (10.14)$$

since destructive interference from times of opposing phase will be missing. For this reason, the electron excitation, and capture and loss cross sections of atoms when bombarded by other atoms or ions, show very sharp maxima when the incident particle velocity corresponds to the electron orbital velocity, as will be seen in the next section. An even more striking example is the sharp resonance in the photoelectric effect when the frequency of the incident radiation is given by the Bohr frequency condition,

$$\hbar\omega = E_{n_0} - E_k \qquad (10.15)$$

If H'_{kn_0} is nearly constant over a period of the exponential, it may be taken out from under the integral sign and Eq. (10.13) can be written after a remaining trivial integration over the time

$$a_k(t) = H'_{kn_0} \frac{1 - \exp\{-(i/\hbar)(E_{n_0} - E_k)t\}}{E_{n_0} - E_k}, \qquad k \neq n_0 \qquad (10.16)$$

The probability of transition to the state k, that is, the probability of finding the system in the state k after a time t, is obtained from $a_k^*(t)a_k(t)$,

$$a_k^*(t)a_k(t) = |H'_{kn_0}|^2 \, 2\left[\frac{1 - \cos(E_{n_0} - E_k)(t/\hbar)}{(E_{n_0} - E_k)^2}\right] \qquad (10.17)$$

Note that if H'_{kn_0} is completely independent of the time (i.e., if the disturbance acts for an infinite time), then since the time or phase average of $1 - \cos x = (2\sin^2 x/2)$ is $1/2$, the same result is obtained as in time-independent perturbation theory, Eq. (9.15).

Problem 10.1 What is the probability of finding the physical system (atom, molecule, or any other sub-microscopic physical entity) in the state n_0 after a time t seconds? How does this compare with the wave function that would be obtained with the time-independent perturbation theory of Chapter 9? Discuss. (Hint: Integrate Eq. (10.10) over the time with $k = n_0$ and $a_{n_0}(0) = 1$.)

10.2 EXCITATION OF ATOMS BY ION BOMBARDMENT

As an illustration of the time-dependent perturbation theory we will discuss atomic excitation by alpha particles (He^+ ions), as first treated by J. A. Gaunt. In this example an atomic electron is momentarily struck by an ion whose effect can be represented by the time-dependent perturbation potential

$$V = -Ze^2/r \tag{10.18}$$

where Z is the charge of the incident ion and r, the distance of the incident ion from an atomic electron, is given in terms of the electron coordinates in the plane of impact, x and y, as shown in Figure 10.1. Only a distant encounter of the alpha particle and the atomic electron is considered here; that is, $x^2 + y^2 \ll b^2$ where b represents the distance of closest approach (the perpendicular distance of the ion trajectory from the target atom). Let v be the ion velocity, t the time measured from the moment of closest approach, and $r_0^2 = b^2 + (vt)^2$; then if x and y are always much less than r,

$$r = [(b - x)^2 + (vt - y)^2]^{1/2}$$

$$\sim [b^2 + (vt)^2]^{1/2}\left[1 - \frac{xb + yvt}{r_0^2}\right] \tag{10.19}$$

Substituting this potential into Eq. (10.13) and integrating over the ion trajectory, that is, over the time from minus to plus infinity, we obtain

$$a_k = \frac{-Ze^2}{i\hbar} \int_{-\infty}^{\infty} dt \int \psi_k{}^*(\mathbf{r})\left\{\frac{1}{\sqrt{b^2 + (vt)^2}}\left[1 + \frac{xb + yvt}{b^2 + (vt)^2}\right]\right\}\psi_{n_0}\, d\tau$$
$$\times \exp\{-(i/\hbar)(E_{n_0} - E_k)t\} \tag{10.20}$$

where the ψ's are, of course, the atomic electron wave functions, and the space integral is taken over the atomic electron coordinates. Since the ψ's are orthogonal, only the second term in the braces gives a finite result, and hence Eq. (10.20) becomes

$$a_k = \frac{-Ze^2}{i\hbar} \int_{-\infty}^{\infty} dt \int \frac{(\psi_k^* x \psi_{n_0})b + (\psi_k^* y \psi_{n_0})vt}{[b^2 + (vt)^2]^{3/2}} \, d\tau$$

$$\times \exp\left\{-(i/\hbar)(E_{n_0} - E_k)t\right\} \tag{10.21}$$

The rest is a matter of arithmetic and the solution will not be continued. Note that the integrand in (10.21) consists of an imaginary exponential

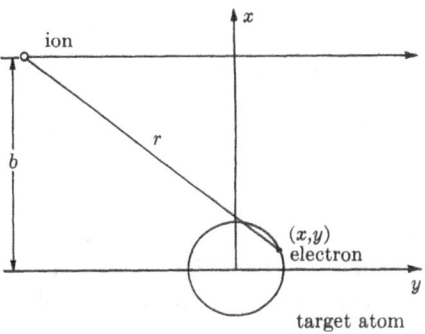

Figure 10.1 *Illustration of collision of positive ion with atomic electron. The ion path deviates very slightly because of collision with the electron.*

function of time and a cumbersome time-dependent coefficient. Note further that when the coefficient of the exponential in the integrand changes appreciably within the period of the exponential oscillation, least cancellation between contributions to the integral of opposite phase occurs. Thus, if the coefficient to the exponential in the integral changes appreciably within the time given by Eq. (10.14), a_k does indeed become a maximum. The period of maximum transition probability is given roughly by the time in which the denominator doubles in magnitude from its minimum value at $t = 0$:

$$T \sim b/v \sim h(E_{n_0} - E_k)^{-1} \sim h/E_n$$

that is, the period of the atomic electron in its original unexcited state. Since a_k is proportional to $1/b^2$, the transition probability increases as b decreases down to b = orbital radius where the approximation fails. Hence, for maximum excitation or ionization probability the incident ion velocity should correspond to b/T where b is of the order of the atomic radius and T is the period of the electron ground state wave function, $2\pi b/v_e$. Phrased differently, maximum atomic excitation results when the alpha particle velocity is of the order of the classical electron orbital velocity.

Problem 10.2   By integrating Eq. (10.21) over the time, estimate the conditions under which the x and the y terms are most important to a transition.

10.3   ABSORPTION OF RADIATION

The next matter to be discussed is the transitions induced by incident electromagnetic radiation, that is, the probability of absorption of a photon. For optical wavelengths (thousands of Angstroms), the incident radiation consists of a time-varying electric field vector whose wavelength is so much greater than atomic dimensions that the electric field may be regarded as constant in space. The x component of the incident electric field vector may be written

$$E_x = E_{0x}(e^{i\omega t} + e^{-i\omega t}) \tag{10.22}$$

and the perturbation potential energy is given by

$$V = E_x \sum e_j x_j \tag{10.23}$$

where x_j is the x displacement of the atomic electron j in the electric field and the summation must be carried out over all the electrons in the atom. This perturbing potential is then substituted into Eq. (10.13), to yield

$$a_k(t) = \frac{E_{0x}}{i\hbar} \sum_j e_j \int_0^t \left(\int \psi_k x_j \psi_{n_0} d\tau \right)$$

$$\times \exp\{-(i/\hbar)(E_{n_0} - E_k)t'\}(e^{i\omega t'} + e^{-i\omega t'}) dt' \tag{10.24}$$

The sum over j of e_j times the space integral is the x component of the dipole moment of the system, which we shall represent by μ, so that the x component of μ taken between the states k and n_0 will be

$$\mu_{x_{kn_0}} = \sum_j e_j \int \psi_k^*(x_j)\psi_{n_0} \, d\tau \qquad (10.25)$$

Performing the indicated integration of (10.24) over the time, we obtain

$$a_k(t) = E_{0x}\mu_{kn_0}\left\{\frac{1 - \exp\left\{-(i/\hbar)(E_{n_0} - E_k + \hbar\omega)t\right\}}{E_{n_0} - E_k + \hbar\omega}\right.$$

$$\left. + \frac{1 - \exp\left\{-(i/\hbar)(E_n - E_k - \hbar\omega)t\right\}}{E_{n_0} - E_k - \hbar\omega}\right\} \qquad (10.26)$$

Only the first term in this expression yields an appreciable result, and that only for frequencies ω close to $(E_k - E_{n_0})/\hbar$, at which the denominator becomes infinitesimal. (n_0 is assumed to be the ground state.) Since

$$\left|\exp\left\{-(i/\hbar)(E_{n_0} - E_k + \hbar\omega)t\right\}\right.$$

$$\left.\times \left[\exp\left\{(i/\hbar)(E_{n_0} - E_k + \hbar\omega)t\right\} - \exp\left\{-(i/\hbar)(E_{n_0} - E_k + \hbar\omega)t\right\}\right]\right|^2$$

$$= 4\sin^2[(1/2\hbar)(E_{n_0} - E_k + \hbar\omega)t]$$

therefore

$$|a_k(t)|^2 = 4(\mu_{x_{kn_0}})^2 E_{0x}^2 \frac{\sin^2[(1/2\hbar)(E_{n_0} - E_k + \hbar\omega)t]}{(E_{n_0} - E_k + \hbar\omega)^2} \qquad (10.27)$$

Equation (10.27) is the probability as a function of time for electromagnetic radiation (of frequency ω, intensity $2E_{0x}^2$, and polarization along the x axis) to be absorbed by an atom, which then undergoes a transition from a state n_0 to a state k as a result of the absorption.

In many cases of interest not involving resonance effects, when the incident radiation is not bright-line spectra (say from the same atoms or nuclei as are absorbing the radiation), the intensity of the incident radiation is essentially a constant function of frequency compared with the frequency dependence of the absorption function (Eq. 10.27). Under these circumstances, Eq. (10.27) must be integrated over all the frequencies contained in the incident radiation. The right side of Eq. (10.27) is negligible for all frequencies except that for which the denominator becomes zero, and therefore the integral may as well extend from zero to infinite frequency. We may note that the frequency-dependent part of (10.27) is essentially a Dirac delta function $\delta(E_{n_0} - E_k + \hbar\omega)$, for

$$\int_0^\infty \frac{\sin^2[\hbar\omega - (E_k - E_{n_0})]t/2\hbar}{[\hbar\omega - (E_k - E_{n_0})]^2} d\omega = \frac{t}{2\hbar^2} \int_0^\infty \frac{\sin^2(x - x_0)}{(x - x_0)^2} dx = \frac{\pi t}{2\hbar^2}$$

where $x_0 = (E_k - E_{n_0})t/2\hbar$. Hence integration of Eq. (10.27) over all frequencies would yield the following result for a frequency continuum of incident radiation:

$$|a_k(t)|^2 = \frac{\pi}{\hbar^2}(\mu_{x_{k n_0}})^2 E_{0x}^2(\omega_{k n_0})t \tag{10.28}$$

where $\omega_{k n_0}$ is the frequency for which the denominator in Eq. (10.27) is zero and $E_{0x}(\omega_{k n_0})$ is the amplitude of the incident electric vector at that frequency. Note that because of the integration over the frequency the transition probability is linearly proportional to the time, for short times, instead of quadratically dependent on time as in Eqs. (10.17, 10.27). Given the intensity of the incident radiation, it is trivial to compute the transition probability per unit time from Eq. (10.28), once the interaction integrals over all space, involving the original and final state wave functions, are known. We will now compute these integrals.

10.4 GENERAL SELECTION RULES FOR SPHERICAL HARMONIC WAVE FUNCTIONS

In Chapter 8 the general solution to a wave equation for a system with a spherically symmetric potential was shown to be expressible in product wave functions of the separated coordinate variables. For the important class of problems for which the unperturbed potential function is spherically symmetric it is well to express the dipole moment in spherical coordinates. The transformation from Cartesian to spherical coordinates is readily made.

The x, y, z components of the dipole moment operator are

$$\mu_x = \mu(r) \sin \theta \cos \varphi$$
$$\mu_y = \mu(r) \sin \theta \sin \varphi \qquad (10.29)$$
$$\mu_z = \mu(r) \cos \theta$$

The dipole moment may result from radiation, an incident alpha particle, or other sources of perturbing potentials. $\mu(r)$ is a function or r alone, say er in the simple case of an atomic electron at radius r from the center of positive charge.

Hence the x component of the dipole moment between two spherical harmonic wave functions becomes

$$\mu_{x_{0,f}} = \iiint \{ R_{n_f l_f}{}^*(r) \, \Theta_{l_f m_f}{}^*(\theta) \, \Phi_{m_f}{}^*(\varphi) \, \mu(r) \sin \theta \cos \varphi$$
$$\times R_{n_0 l_0}(r) \, \Theta_{l_0 m_0}(\theta) \, \Phi_{m_0}(\varphi) \, r^2 \sin \theta \} \, dr \, d\theta \, d\varphi \qquad (10.30)$$

where the subscript on the left-hand member has been abbreviated to 0, f for original and final states. The letter n is the total quantum number for hydrogen wave functions or the radial quantum number for most other wave functions. The orbital quantum number l and the magnetic quantum number m complete the three eigenvalues needed to completely specify a three-dimensional wave function. The three integrations may be rewritten as a product of three separate integrals. For the wave function in φ, Eq. (8.6), one obtains an integral over φ for μ_x, for example,

$$\frac{1}{\pi} \int_0^{2\pi} e^{-i m_f \varphi} \left(\frac{e^{i\varphi} + e^{-i\varphi}}{2} \right) e^{i m_0 \varphi} \, d\varphi \qquad (10.31)$$

since $\cos \varphi = (e^{i\varphi} + e^{-i\varphi})/2$.

The integral of $e^{-(m_f - m_0)\varphi}$ is zero unless $m_f - 1 - m_0 = 0$, when its value is unity. Including the integral of $e^{-i\varphi}$, the entire integral over φ in Eq. (10.31) will be zero unless

$$m_f = m_0 \pm 1 \qquad (10.32)$$

in which cases it is unity. Thus we obtain our first selection rule, Eq. (10.32), giving the changes in magnetic quantum number permitted when the incident field is polarized in the x–direction. For incident field polarized in the z-direction the resultant selection rule for μ_z would be $m_f = m_0$, i.e., no change in m is possible. This is understandable if one realizes that a force in the z direction can have no effect on the motion in the xy plane which is responsible for the component of the angular momentum along the z axis represented by the m, or magnetic, quantum number. A dipole transition may result from a passing ion, radiation, or any other perturbing potential. We have obtained the universal and important result that the magnetic quantum number can change by no more than one unit in a dipole transition between unperturbed wave functions resulting from spherically symmetric potentials. That is,

$$m_f = m_0 \quad \text{or} \quad m_0 \pm 1 \tag{10.32a}$$

In a similar way selection rules for the l quantum number may be obtained from the integral over θ. For μ_z, for example, and $m = 0$, for which the Θ functions are Legendre polynomials, the θ integral in (10.30) would be

$$\int_0^1 P_{l_f}(x) x P_{l_0}(x) \, dx \tag{10.33}$$

where $x = \cos \theta$. Since the Legendre polynomials are orthogonal and x times a polynomial is a different polynomial, l_f must be different from l_0 in order for a finite result to be obtained. Indeed, it is a property of the Legendre polynomials that they satisfy the relation

$$x P_l(x) = \frac{(l + 1)P_{l+1}(x) + l P_{l-1}(x)}{2l + 1} \tag{10.34}$$

so that Eq. (10.33) is zero unless

$$l_f = l_0 \pm 1 \tag{10.35}$$

in which cases the integral (10.33) would equal either $(l + 1)/(2l + 1)$ or $l/(2l + 1)$. When $m \neq 0$ and the associated Legendre polynomials must be used, the same selection rule for l is obtained. That is, *in a dipole transition the orbital angular momentum must change by one unit for a one-electron wave function.*

Total angular momentum is conserved as it must be always, because a photon has an intrinsic spin of unity. Photons may be represented as a superposition of circularly polarized electromagnetic waves each one of which always has one unit of angular momentum along its direction of propagation, positive or negative according to whether the light is right or left circularly polarized. It is of interest to note that such an effect can be true only for massless motion (photon or neutrino) with the speed of light. For slower particles one can always transform to a velocity frame in which the particle's motion is reversed, in which case angular momentum cannot be conserved if the particle's intrinsic spin or angular momentum is quantized along its direction of propagation.

For matrix elements different from the dipole transition expressed in Eq. (10.25), the light quantum may have additional units of angular momentum; aside from its intrinsic angular momentum, its wave function may possess orbital momentum about the origin in non-dipole matrix elements such as quadrupole or octupole transitions. In general, the electric field of an incident plane wave would be given by

$$E_x = E_{0x}e^{i(kz - \omega t)} \tag{10.22a}$$

where k is the wave number of the incident photons. A plane wave e^{ikz} may be expanded in terms of the complete orthonormal set of spherical harmonic wave functions:

$$e^{ikz} = e^{ikr\cos\alpha} = \sum_{n=0}^{\infty} (2n + 1)i^n \sqrt{\frac{\pi}{2kr}} \, J_{n+\frac{1}{2}}(kr) \, P_n(\cos\alpha) \tag{10.36}$$

where $J_{n+\frac{1}{2}}$ is a Bessel function of order $n + \frac{1}{2}$ and α is measured from the propagation vector of the plane wave, *NOT necessarily from the atomic coordinate axis* as, for example, in Eq. (10.29). If Eqs. (10.36) and (10.22a) are substituted into (10.23) and the computation repeated, we will obtain elements other than the dipole transition matrix elements for $n \neq 0$. Near the origin

$$J_{n+\frac{1}{2}}(kr) \sim \frac{1}{(n + \frac{1}{2})!}\left(\frac{kr}{2}\right)^{n+\frac{1}{2}}$$

Thus the expansion yields a quadrupole moment r^2, an octupole moment r^3, and so on for the higher order terms. In virtually all electromagnetic transitions (nuclear or atomic), $kr < < 1$ for r less than the nuclear or atomic radius, which makes the expansion particularly useful since it converges rapidly.

The only remaining integral is the radial integral and that is a function of the radial dependence of the perturbing potential and the unperturbed radial wave functions. Hence no general selection rules or other results can be obtained for the radial integrals. Selection rules for the special cases of the harmonic oscillator and the hydrogen atom will be discussed in the next two sections.

10.5 SELECTION RULES FOR THE HARMONIC OSCILLATOR RADIAL QUANTUM NUMBER

The harmonic oscillator is of great physical interest since it represents electromagnetic radiation confined in a box (nucleons in a nucleus, and diatomic molecular vibrations). The dipole moment between different harmonic oscillator wave functions yields the probability of transitions between nuclear states of different energy, or of a light quantum being added to or subtracted from the total radiation in a box. The radial harmonic oscillator wave functions, Eqs. (8.49, 8.51b), are comprised in part of an even or odd power series in r, the even or odd spherical Hermite polynomials. Similar to the case of the associated Legendre polynomials, the radial integral yields the following selection rules:

$$
\begin{aligned}
n_f &= n_0, \ n_0 - 1 \quad \text{for} \quad l_f = l_0 + 1 \\
n_f &= n_0 + 1, \ n_0 \quad \text{for} \quad l_f = l_0 - 1
\end{aligned}
\tag{10.37}
$$

For light waves in a rectangular box, the one-dimensional harmonic oscillator integral gives

$$
n_f = n_0 \pm 1
$$

corresponding to the creation of one quantum in addition to those already contained in the box, or to the annihilation or loss of a quantum due to some absorption process. For a nucleon or a material oscillator representing a light wave, the wave equation in one dimension is

$$
\frac{\partial^2 \psi}{\partial x^2} + \frac{2m}{\hbar^2}\left(E - \frac{m}{2}\omega_0^2 x^2\right)\psi = 0
\tag{10.38}
$$

whose solution is

$$
\psi_n = A_n H_n(\sqrt{m\omega_0/\hbar}\,x)\,e^{-m\omega_0 x^2/2\hbar}
\tag{10.39}
$$

where H_n is the Hermite polynomial of order n and A_n is a normalization constant. The energy eigenvalue of the nth wave function is

$$E_n = (n + \tfrac{1}{2})\hbar\omega_0 \tag{10.40}$$

which may be interpreted as meaning that the material harmonic oscillator of frequency ω_0 has n quanta of energy. If a photon is emitted, one quantum is lost from the oscillator and the energy of the new state is $(n - \tfrac{1}{2})\hbar\omega_0$, the eigenvalue of the ψ_{n-1} wave function. The harmonic oscillator is a convenient way to describe the radiation field, since the energy of the radiation field is also given by $E_n = n\hbar\omega_0$. Of course the photon wave functions are quite different from the material oscillator wave functions, since they are solutions to a different wave equation.

10.6 THE HYDROGEN ATOM

Pauli evaluated some of the radial dipole moment integrals for the hydrogen atom. From the previous discussion of the spherical integrals it was learned that all spherically symmetric potentials result in the selection rules Eqs. (10.32a, 10.35) for the total orbital and magnetic quantum numbers. By use of the wave functions for the electron in the hydrogen atom, Eqs. (8.38, 8.45), the radial electric dipole moment integral was evaluated. For $l_0 = 0$, $n_0 = 1$ (the Lyman series) ($l_f = 1$ in accordance with the selection rule),

$$\mu_{n_f n_0} = \int R_{n_f l_f}{}^*(r) r R_{n_0 l_0}(r) r^2 \, dr$$

$$= \frac{(n_f - 1)^{2n_f - 1}}{n_f (n_f + 1)^{2n_f + 1}} \tag{10.41}$$

and similarly, for $l_0 = 0$ or 1 ($l_f = 1$ or $0,2$) and $n_0 = 2$ (the Balmer series),

$$\mu_{n_f n_0} = \frac{\cdot \, (n_f - 2)^{2n_f - 3}}{n_f (n_f + 2)^{2n_f + 3}} (3n_f{}^2 - 4)(5n_f{}^2 - 4) \tag{10.42}$$

As can be seen from Eqs. (10.41, 10.42), no selection rule results for the radial quantum number. Thus dipole transitions are possible from any radial quantum number to any other radial quantum number, if l changes by only one unit and m changes by no more than one unit. The relative absorption probabilities for different transitions are given by the ratios of the squares of Eqs. (10.41) and (10.42). The Lyman spectra absorption probability decreases rapidly as n_f increases, behaving roughly as n_f^{-6}.

Problem 10.3 Find the probability of transition of a hydrogen atom from the ground state to the state $n,l,m = 2,1,1$ if an alpha particle passes with velocity v at a distance b from the hydrogen nucleus.

Problem 10.4 The potential for a neutron interacting with an electron might be represented by an attractive spherical square well 10^4 ev deep and 10^{+12} cm in radius. If a neutron interacts for 10^{-17} second with a hydrogen electron in its ground state, what are the probabilities of its causing transitions to the first three excited states? (Assume the neutron is located at the center of the atom during this time.)

Problem 10.5 If an electric field of 10^4 volts/m acts for 10^{-17} second on a hydrogen atom in its ground state, what is the probability of finding the atom subsequently in a $2s$ or $2p$ state?

10.7 EMISSION

So far all the theoretical development in this chapter pertains strictly to absorption. We would have to use a more difficult and sophisticated procedure for direct calculation of emission; instead, we will show by a thermodynamic argument that emission and absorption are equivalent except for a simple coefficient, and thus show how the emission may be calculated in terms of the absorption. The procedure will be to find the emission in terms of the absorption.

Let the ground state of a radiating and absorbing system be labeled i and an excited state k. The rate of absorption is linearly proportional to the density of incident radiation $u(\nu)$ times the number of atoms in the ground state, N_i,

$$\text{Absorption rate} = Bu(\nu)N_i \tag{10.43}$$

where B is a constant known as the *Einstein B* coefficient.

We have calculated the absorption of radiation in Section 10.3. Since E_{0x} occurring in Section 10.3 is only one of three components of the radiation in a box (E_{0y}, E_{0z} being the others) $E_{0x}{}^2 = u(\nu)/3$. Equation (10.28) gives the transition probability for one atom in time t seconds. Hence the absorption rate per unit volume is

$$| a_k(t) |^2 N_i/t = \frac{\pi}{3\hbar^2}(\mu_{ki})^2 u(\nu)N_i \tag{10.43a}$$

Comparison of (10.43a) with (10.43) yields

$$B = \frac{\pi}{3\hbar^2} | \mu_{mn} |^2 \tag{10.44}$$

It is convenient to divide the emission process into two parts, the spontaneous emission which occurs independently of the environment and the induced emission resulting when the excited system is in a bath of radiation:

$$\text{Emission rate} = AN_k + Cu(\nu)N_k \tag{10.45}$$

where A is the *Einstein A* coefficient for spontaneous emission and C is the coefficient for induced emission. The induced emission results from the second term in Eq. (10.26), and therefore it should have the same constant coefficient as the absorption resulting from the first term in Eq. (10.26). We shall soon discover by the present thermodynamic argument that this is indeed the case.

At equilibrium the absorption and emission rates must be equal, that is,

$$Bu(\nu)N_i = AN_k + Cu(\nu)N_k$$

whence

$$\frac{A}{Bu(\nu)} + \frac{C}{B} = \frac{N_i}{N_k} \tag{10.46}$$

The number of systems in a state having energy E is proportional to $e^{-E/kT}$, a result borrowed from statistical mechanics which we used in Chapter 1 to derive Eq. (1.13). Therefore, the ratio of unexcited atoms to excited atoms, occurring on the right-hand side of (10.46), is

$$\frac{N_i}{N_k} = e^{-(E_i - E_k)/kT} = e^{h\nu/kT} \tag{10.47}$$

where ν is the frequency of the absorbed or emitted radiation. $u(\nu)$ which also occurs in (10.46) is given by Planck's formula, Eq. (1.21),

$$u(\nu)\,d\nu = \frac{8\pi h\nu^3/c^3}{e^{h\nu/kT} - 1}\,d\nu \tag{1.21}$$

Thus Eq. (10.46) becomes

$$\frac{c^3}{8\pi h\nu^3}(e^{h\nu/kT} - 1)\frac{A}{B} + \frac{C}{B} = e^{h\nu/kT} \tag{10.48}$$

This equation holds at any temperature or frequency. Since it must hold at $T = \infty$, $C = B$, and since it must hold at $T = 0$,

$$A = \frac{8\pi h\nu^3}{c^3}B \tag{10.49}$$

Thus, we do not need to rely on quantum electrodynamics, which is beyond the scope of this text, to calculate the spontaneous emission of an excited oscillator; we may find the spontaneous emission from the simple absorption calculations given in this chapter. For example, the lifetime for an excited state pending a dipole transition is given by

$$\tau = \frac{1}{A} = \frac{3\hbar c^3}{4\omega^3}|\,\mu_{mn}\,|^{-2} \tag{10.50}$$

Problem 10.6   What is the lifetime of a hydrogen atom in the $(n = 2, l = 1, m = 0)$ state? In the $(n = 3, l = 2, m = 0)$ state?

Problem 10.7   A diatomic molecule vibrates approximately like a harmonic oscillator, with a fundamental frequency of the order of 10^{10} cycles per second. If diatomic molecules are inside a resonant microwave cavity of the same frequency in the (1,1,0) mode and $E = E_z = 10^4$ volts/m, what is the rate of transition between the ground and first excited states? What fraction of the molecules are in an excited state?

10.8 SECOND-ORDER TIME-DEPENDENT PERTURBATION THEORY

As in time-independent perturbation theory, the first-order time-dependent theory involving a direct transition from an initial state n_0 to a final state k may give a null result. That is, we may find $H'_{kn_0} = 0$. Under these circumstances it is necessary to multiply Eq. (10.9) by Ψ_j^* on the left where Ψ_j^* represents any state that may be reached by a direct transition. Thus we will have $H'_{jn_0} \neq 0$ and, if H'_{jn_0} is nearly constant over a period of the exponential, we will obtain an amplitude for transition to the state Ψ_j given by

$$a_j(t) = H'_{jn_0} \frac{1 - \exp\{-(i/\hbar)(E_{n_0} - E_j)t\}}{E_{n_0} - E_j}, \qquad j \neq n_0 \qquad (10.16)$$

The total wave function now becomes (as before)

$$\Psi = \sum_j a_j(t)\Psi_j$$

$$\Psi = \Psi_{n_0} + \sum_{j \neq n_0} H'_{jn_0} \frac{1 - \exp\{-(i/\hbar)(E_{n_0} - E_j)t\}}{E_{n_0} - E_j}\Psi_j \qquad (10.51)$$

where in the first term $a_{n_0} = 1$. Equation (10.51) may be used as an initial wave function in Eq. (10.11). This amounts to saying that for $j \neq n_0$ the $a_j(t)$ in Eq. (10.10) are $\neq 0$ but are given by Eq. (10.16). Hence, by repeating the calculation that gave us Eq. (10.16), we obtain

$$i\hbar\dot{a}_k(t) = \sum_j a_j(t)H'_{kj}\, e^{-(i/\hbar)(E_j - E_k)t} \qquad (10.10)$$

$$= H'_{kn_0} \exp\{-(i/\hbar)(E_{n_0} - E_k)t\}$$

$$+ \sum_j \frac{H'_{jn_0}H'_{kj}}{E_{n_0} - E_j}[1 - \exp\{-(i/\hbar)(E_{n_0} - E_j)t\}]\exp\{-(i/\hbar)(E_j - E_k)t\}$$

$$(10.52)$$

and

$$a_k(t) = H'_{kn_0} \frac{1 - \exp\{-(i/\hbar)(E_{n_0} - E_k)t\}}{E_{n_0} - E_k}$$

$$+ \sum \frac{H'_{jn_0} H'_{kj}}{E_{n_0} - E_j}$$

$$\times \left[\frac{1 - \exp\{-(i/\hbar)(E_j - E_k)t\}}{E_j - E_k} - \frac{1 - \exp\{-(i/\hbar)(E_{n_0} - E_k)t\}}{E_{n_0} - E_k}\right]$$

$$(10.53)$$

The terms in $E_j - E_k$ resulting from the first term in brackets may be neglected for the usual case of $|E_j - E_k| \gg |E_{n_0} - E_k|$ (this is because if $|E_{n_0} - E_k|$ is not very small, the $a_k(t)$ are negligible); otherwise, they must be included. If the first term of (10.53) is zero due to the vanishing of H'_{kn_0}, the transition probability then becomes

$$|a_k(t)|^2 = \left|2\sum_j \frac{H'_{kj} H'_{jn_0}}{E_{n_0} - E_j}\right|^2 \frac{1 - \cos\left[(E_{n_0} - E_k)(t/\hbar)\right]}{(E_{n_0} - E_k)^2} \qquad (10.54)$$

In second-order time-dependent perturbation theory, just as in time-independent perturbation theory, occupation of the final state occurs through available intermediate states whose transition probabilities must be summed over. A curious result is that the transition amplitude to a particular state depends critically on the presence of other states whose energy is nearly the same. The initial state is, as usual, a pure state ($a_i = \delta_{in_0}$), but in second-order theory, transitions to the final state occur through intermediate states which, although initially vacant, have a small but finite probability in first-order perturbation theory of becoming populated.

When the perturbation is due to continuous incident radiation (that is, when the light intensity development leading to Eq. (10.28) is valid), we obtain for second-order perturbation theory

$$|a_k(t)|^2 = \frac{1}{\hbar^2}\left|\sum_{j \neq k, n_0} \frac{\mu_{x_{kj}} \mu_{x_{jn_0}}}{E_{n_0} - E_j}\right|^2 E_{0x}^2(\omega_{kn_0})t \qquad (10.55)$$

which is analogous to Eq. (10.28). Equation (10.54) may be generalized to yield the third-order correction,

$$|a_k(t)|^2 = 2\left|\sum_j \sum_i \frac{H'_{kj}H'_{ji}H'_{in_0}}{(E_{n_0} - E_j)(E_{n_0} - E_i)}\right|^2 \frac{1 - \cos\left[(E_{n_0} - E_k)(t/\hbar)\right]}{(E_{n_0} - E_k)^2}$$

which includes transitions via two intermediate states.

10.9 TRANSITIONS TO OR FROM UNBOUND SYSTEMS

Both light emission and absorption are calculated by use of matrix elements (10.11) H'_{kn_0}, which are the same for both processes since the processes are similar. It may be, though, that either the final state of the particle is unbound, as would be the case in atomic ionization due to the photoelectric effect, or the initial state is unbound as occurs in light emission caused by electron capture. *If either the initial or final particle state is free, the particle energy spectrum in this state is no longer discrete,* and given continuous radiation we no longer observe or are interested in finding the transition probability to or from a particular state but instead concern ourselves with an energy interval between E and $E + dE$ and assume that all states within this interval have equal probability of being occupied. The probability of a transition occurring to some state in this energy interval will be proportional to the number of states in the interval, $\rho(E)\,dE$, where $\rho(E)$ is the level density or number of states per unit energy interval. In the case of a free electron or nucleon where both spin polarizations must be included,

$$\rho(E)dE = p^2\,dp\,\frac{V}{(\hbar\pi)^3} \tag{10.56}$$

from Eq. (1.12), for example, where p is the momentum of the free particle and V is the volume in which the free particle wave functions are contained. Equation (1.12) for radiation in a box applies to free particles in a box, as we know, since the particles in a box have a probability wave amplitude which satisfies a wave equation similar to the equation for the the photon wave function, Eq. (1.5). The boundary conditions are also similar and even the twofold degeneracy due to particle spin values of $\pm\frac{1}{2}$ is matched by the two degrees of photon transverse polarization freedom. Thus, for a free electron or nucleon the level density obeys the proportionality

$$\rho(E) \, dE \propto p^2 \, dp \propto \sqrt{E} \, dE \qquad (10.56a)$$

While the level density is directly proportional to the volume in which the particle is contained, the matrix elements, on which the transition probability also depends (Eq. 10.57 below), are directly proportional to the normalization constant of the free particle wave function, C_N. Since the square of the matrix elements is used to obtain the transition probability, the transition probability is proportional to the square of the normalization constant for the free particle wave function. The normalization constant is so defined that $|C_N|^2 \int \psi^* \psi \, d\tau = 1$ with the integral extending over the volume of the box in which the particle moves. Since for free particle wave functions $\psi = e^{i\mathbf{k} \cdot \mathbf{r}}$ the integral is equal to this volume, $|C_N|^2 = 1/V$ and hence the transition probability $\propto \rho(E)|C_N|^2$ is independent of the volume. For convenience, in the following we will assume the free particle contained in a unit volume.

The probability of transition of a particle to or from a free state by absorption or emission of a photon of frequency ω is obtained by integrating the corresponding expression for discrete initial and final states (and for continuous ω instead of continuous E), Eq. (10.27), over all the equally probable states with energies E (in the case of the *initial* state being free but unknown an *average* is taken); in the case of absorption by a system in state n_0 to any of many equally probable states in the energy interval dE,

$$|a(E)|^2 = \int 4|H'_{kn_0}|^2 I(\omega) \left\{ \frac{\sin^2\left[(E_{n_0} - E + \hbar w)(t/2\hbar)\right]}{(E_{n_0} - E + \hbar w)^2} \right\} \rho(E) \, dE \qquad (10.57)$$

where the general matrix element H'_{kn_0} has been written instead of the particular matrix element for dipole transitions, μ_{kn_0}. $I(\omega)$ represents the intensity of radiation of frequency ω, n_0 labels the discrete energy level (bound) state, and k any continuous (unbound) state of energy E. Note that Eq. (10.27) was derived by assuming (as in obtaining Eq. 10.12) that the perturbation is small and does not deplete the initial state appreciably; that is, $a_{n_0} = 1$ always. As in the derivation of Eq. (10.28), the quantity in braces behaves like a delta function for $t \gg \hbar/(E_{n_0} + \hbar\omega)$, so that Eq. (10.57) becomes upon integration

$$|a(E)|^2 = (2\pi/\hbar)I(w)\rho(E_{n_0} + \hbar w)|H'_{kn_0}|^2 t$$

The transition probability per unit time between states of energy E_{n_0} and $E_{n_0} + \hbar\omega$ is a constant given by

$$T(E_{n_0} + \hbar\omega, E_{n_0}) = (2\pi/\hbar)I(\omega)\rho(E_{n_0} + \hbar\omega)|H'_{kn_0}|^2 \qquad (10.58)$$

As t becomes very large the initial state is depleted so that the initial condition, $a_{n_0} \sim 1$, is no longer valid, and the theory given here based on this simple condition breaks down.

Problem 10.8   What energy dependence of the transition rate is to be expected for neutrons emitted in a photonuclear transition if H'_{kn_0} is independent of energy and the nucleus is left in its ground state?

Problem 10.9   What is the rate of ionization of hydrogen atoms in the ground state when in light of which the intensity I_0 is constant over all frequencies up to twice the minimum ionization frequency?

Problem 10.10   Fermi assumed that in some circumstances the beta-decay transition matrix H'_{kn_0} is independent of electron energy and that two particles, an electron and a neutrino, are emitted in beta decay. The sum of the electron and neutrino energy must be a constant and the sum of the momenta of the recoil nucleus, electron, and neutrino must be zero in the laboratory system. Derive the beta-particle energy spectrum resulting from these assumptions.

Problem 10.11   If in a sample of tritium H^3 there are 10^{19} tritium atoms per cubic centimeter, how many beta-decay formed He^+ ions in the $(n = 3, l = 1)$ excited state are there per cubic centimeter? (See Problem 9.12.) (Hint: Compare the observed half-life of H^3, 12.26 yrs with the atomic deexcitation rate of He^+ in this state and the fraction of H^3 which decays initially into this state.)

BIBLIOGRAPHY

Pauling, L., and E. B. Wilson, cited in Chapter 7.

Heitler, W., *The Quantum Theory of Radiation*. London: Oxford University Press, 1950.

CHEMICAL RESONANCE THEORY

11

A SYSTEM COMPOSED of two identical particles, say the two electrons in a helium atom, or a hydrogen molecule, may be described in terms of the superposition of states of the hydrogenic system, the interaction energy between the two particles being treated as a perturbation energy, as in Chapter 9. Perturbation theory tells us how much time the two particles spend in different states of the unperturbed system. An enlightening physical interpretation of this situation may be given if we compare the system to the familiar phenomenon of two weakly coupled oscillators. Two pendula, for example, A and B, may be suspended from the same supporting rod. There are two different coordinate systems in which to represent their motion, a coordinate system in which the motion of each bob is described separately and a combined coordinate system in which no attempt is made to describe their motion separately. After bob A is set swinging, its oscillation will die down and bob B will begin to swing, and so on, and the system will be said, classically, to *resonate* between one state (of the unperturbed system) in which A swings and B is at rest and the other state in which B swings and A rests. Alternatively, one could describe the system consisting of bobs A and B in terms of normal coordinates and say that it is oscillating in two different states (these are said to be the true states of the combined system), with two slightly different characteristic frequencies: One might regard as the ground normal state the one with the lower frequency, in which both bobs swing together with equal amplitude and phase; when the two bobs swing with equal amplitude but with a phase difference of 180° their frequency is slightly higher. If one attempted to represent the ground normal state of the combined system in terms of the other coordinate system, that is, in terms of the motion of either bob, considered as an unperturbed, independently moving pendulum, one would then say that the system was in one of its two unperturbed states, with both bobs swinging with equal amplitudes but with slightly lower frequency than if either bob were swinging completely independent of the motion of the other.

The preceding analogy is of considerable utility in acquiring an insight into chemical bonding. On the one hand an ionic bond, as between sodium and chlorine in a salt molecule, is easily understood. Resonance is not involved. The valence electron of one atom joins the valence electrons of another atom, essentially causing the first atom to be a positive ion and the other a negative ion, and the two ions are held together by simple electrostatic attraction. A covalent bond, on the other hand, as between two hydrogen atoms in a gas molecule, is less easily explained. The electrons may be thought to resonate among all the different degenerate states about the two atoms, causing the total binding energy to be less than the electron binding energy for the two separated hydrogen atoms. Carbon, for example, has four valence electrons and can establish covalent bonds with four nearest carbon neighbors to form the extremely stable diamond crystal. Diamonds, however, are a bad long-term investment since after many eons they decay to a still more stable structure, graphite. In graphite there are only three nearest neighbors in a hexagonal, two-dimensional plane, and the additional resonance energy resulting from the three nearest neighbors sharing the fourth electron for equal times causes the graphite bond to be stronger than the bond in the diamond crystal. Other organic structures of particularly great stability are the aromatic compounds, comprised basically of the hexagonal benzene ring which also has much resonance energy. In general, any chemical bond is due to both the ionic type of bond in which the valence electrons are clustered predominantly about one atom, and the homopolar bond in which the valence electrons remain equally near both nuclei. A complete calculation usually consists of appropriate contributions from both bonds (determined possibly by the variational method of Chapter 9). Exceptions to this are calculations for the sodium chloride bond, which is essentially ionic, and the hydrogen iodide bond, which is essentially homopolar or covalent.

11.1 REVIEW OF CLASSICAL COUPLED OSCILLATORS

If two oscillators free to move in only one dimension are attached to opposite walls through springs of force constant k_1 and connected to each other by a weak spring of force constant k_2, as shown in Figure 11.1, the form of Newton's second law for the two oscillators will be

$$m\ddot{x}_1 = -k_1 x_1 - k_2(x_1 - x_2)$$
$$m\ddot{x}_2 = -k_1 x_2 - k_2(x_2 - x_1)$$

(11.1)

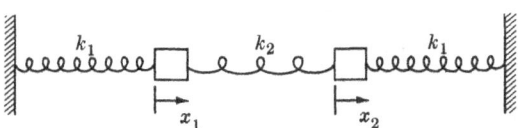

Figure 11.1 *Illustration of two harmonic oscillators connected by a weak spring with constant k_2.*

These equations may be rewritten

$$\ddot{x}_1 = -\omega_0^2 x_1 + \Omega^2 x_2$$
$$\ddot{x}_2 = -\omega_0^2 x_2 + \Omega^2 x_1$$

(11.1a)

where $\omega_0^2 = (k_1 + k_2)/m$ and $\Omega^2 = k_2/m$. These equations take a simpler form if new variables are chosen. Let

$$\xi = \frac{1}{\sqrt{2}}(x_1 + x_2), \qquad \eta = \frac{1}{\sqrt{2}}(x_1 - x_2) \tag{11.2}$$

Then

$$\ddot{\xi} = -\omega_0{}^2\xi + \Omega^2\xi, \qquad \ddot{\eta} = -\omega_0{}^2\eta - \Omega^2\eta \tag{11.3}$$

whose solutions are

$$\xi = A \sin(\sqrt{\omega_0{}^2 - \Omega^2}\, t); \qquad \eta = B \sin(\sqrt{\omega_0{}^2 + \Omega^2}\, t) \tag{11.4}$$

and therefore

$$x_1 = \frac{A}{\sqrt{2}} \sin(\sqrt{\omega_0{}^2 - \Omega^2}\, t) + \frac{B}{\sqrt{2}} \sin(\sqrt{\omega_0{}^2 + \Omega^2}\, t)$$

$$x_2 = \frac{A}{\sqrt{2}} \sin(\sqrt{\omega_0{}^2 - \Omega^2}\, t) - \frac{B}{\sqrt{2}} \sin(\sqrt{\omega_0{}^2 + \Omega^2}\, t) \tag{11.5}$$

At times when $\sqrt{\omega_0{}^2 - \Omega^2}\, t = \sqrt{\omega_0{}^2 + \Omega^2}\, t - 2\pi n$, where n is an integer, x_1 is a maximum and x_2 a minimum. If

$$\sqrt{\omega_0{}^2 - \Omega^2}\, t = \sqrt{\omega_0{}^2 + \Omega^2}\, t - \pi(2n + 1)$$

then x_2 is a maximum and x_1 a minimum. Since $\Omega^2 \ll \omega_0{}^2$,

$$\sqrt{\omega_0{}^2 \mp \Omega^2} \sim \omega_0 \mp (\Omega^2/2\omega_0)$$

and

$$x_1 \sim \frac{A + B}{\sqrt{2}} \cos(\Omega^2/2\omega_0)t \sin \omega_0 t + \frac{B - A}{\sqrt{2}} \sin(\Omega^2/2\omega_0)t \cos \omega_0 t$$

$$x_2 \sim \frac{A - B}{\sqrt{2}} \cos(\Omega^2/2\omega_0 t) \sin \omega_0 t - \frac{B + A}{\sqrt{2}} \sin(\Omega^2/2\omega_0)t \cos \omega_0 t \tag{11.6}$$

If the initial conditions are such that $A = B$,

$$x_1 \sim \sqrt{2}A \cos(\Omega^2/2\omega_0)t \sin \omega_0 t$$

$$x_2 \sim \sqrt{2}A \sin(\Omega^2/2\omega_0)t \cos \omega_0 t \tag{11.7}$$

yielding the two oscillations of frequency ω_0 whose amplitude varies slowly with frequency $\Omega^2/2\omega_0$. These oscillations are illustrated in Figure 11.2. When we watch the springs move we are in a sense analyzing them in the x_1,x_2 coordinate frame where x_1 and x_2 are the coordinates of the two masses on the springs. The motion in the x_1,x_2 coordinate frame is a little complicated: When x_1 is oscillating with maximum amplitude, x_2 is at rest, and vice versa with first one mass oscillating and then the other as the

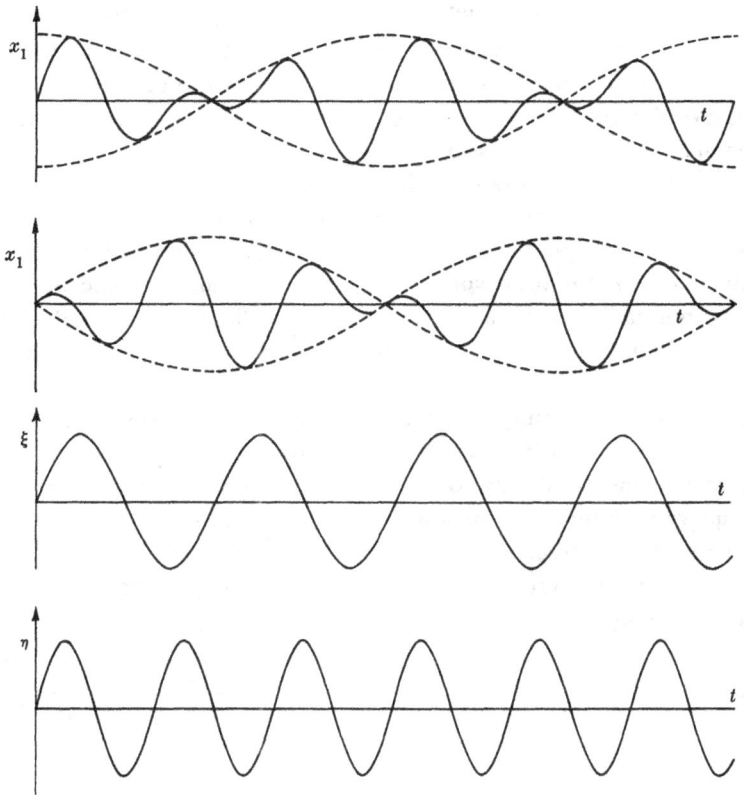

Fig. 11.2 *Illustrations of vibrations of two spring-coupled oscillators in the normal coordinates ξ and η.*

energy of the system is transferred back and forth between the two masses. That is, the system is resonating between the two oscillations in the x_1,x_2 frame; first the system is oscillating in the x_1 state, then in the x_2 state and so on. The analysis of the coupled spring system is very much simplified if we analyze the system (represent it) in the normal coordinate frame, $\xi = (1/\sqrt{2})(x_1 + x_2)$ and $\eta = (1/\sqrt{2})(x_1 - x_2)$. As we can see from Figure 11.2, the motion in normal coordinates is a simple harmonic oscillation of constant amplitude in each coordinate. Added or subtracted together (superimposed) this constant amplitude vibration at two different frequencies results in the complicated x_1, x_2 motion, $x_1 = (1/\sqrt{2})(\xi + \eta)$ and $x_2 = (1/\sqrt{2})(\xi - \eta)$. The initial conditions we picked for the example gave equal amplitudes for the ξ, η motion, but other initial conditions would result in unequal amplitudes, in general, and of course different x_1, x_2 motion than shown here. For example, if both masses received the same initial displacement and velocity, both masses would continue to move in phase forever (if no damping were present; $x_1 = x_2$). Represented in normal coordinates, no η vibration would be present; the motion would be entirely in the ξ normal coordinate. In the x_1,x_2 coordinates both states would occur with equal frequency or probability; in the ξ, η coordinates only one state would occur.

In atomic systems, one would observe directly the frequencies of the ξ,η system, but the analysis would probably be carried out in the unperturbed, independently oscillating x_1,x_2 system. The perturbation from the weak interaction of the two unperturbed degenerate systems is said to result in a resonance energy if the initial wave functions are not the so-called correct zero-order wave functions, thus yielding a nondiagonal perturbation matrix.

11.2 QUANTUM MECHANICAL COUPLED HARMONIC OSCILLATORS

A system of oscillators subject to the conservative forces given on the right side of Eq. (11.1a) has a potential energy given by

$$V = \tfrac{1}{2}\omega_0^2 x_1^2 + \tfrac{1}{2}\omega_0^2 x_2^2 - \Omega^2 x_1 x_2 \tag{11.8}$$

With the substitution of this potential together with (11.2), the Schroedinger equation (4.23) becomes

$$\frac{\partial^2 \psi}{\partial \xi^2} + \frac{\partial^2 \psi}{\partial \eta^2} + \frac{2m}{\hbar^2}\left\{E - \tfrac{1}{2}[(\omega_0^2 + \Omega^2)\xi^2 + (\omega_0^2 - \Omega^2)\eta^2]\right\}\psi = 0 \quad (11.9)$$

The variables are separable: $\psi = \varphi_1(\xi) \cdot \varphi_2(\eta)$; hence

$$\frac{\partial^2 \varphi_1}{\partial \xi^2} + \frac{2m}{\hbar^2}[E_\xi - \tfrac{1}{2}(\omega_0^2 + \Omega^2)\xi^2]\varphi_1 = 0 \quad (11.10a)$$

and

$$\frac{\partial^2 \varphi_2}{\partial \eta^2} + \frac{2m}{\hbar^2}[E_\eta - \tfrac{1}{2}(\omega_0^2 - \Omega^2)\eta^2]\varphi_2 = 0 \quad (11.10b)$$

These equations are the one dimensional harmonic oscillator wave equations whose solutions are the ordinary Hermite functions given at the end of Chapter 8. The energy eigenvalues are given by

$$E_{\xi n} = (n + \tfrac{1}{2})\hbar\sqrt{\omega_0^2 + \Omega^2} \quad (11.11)$$

Similarly,

$$E_{\eta m} = (m + \tfrac{1}{2})\hbar\sqrt{\omega_0^2 - \Omega^2}$$

$$E_{nm} = E_{\xi n} + E_{\eta m} = (n + \tfrac{1}{2})\hbar\sqrt{\omega_0^2 + \Omega^2} + (m + \tfrac{1}{2})\hbar\sqrt{\omega_0^2 - \Omega^2}$$

$$\simeq (n + m + 1)\hbar\omega_0 + \hbar\Omega^2(n - m)/2\omega_0 \quad (11.12)$$

Thus, quantum mechanical coupled oscillators yield basic frequencies $\sqrt{\omega_0^2 \pm \Omega^2}$, the same as the classical system discussed in Section 11.1 (Eq. 11.1). Moreover, if we try to analyze this system in terms of the uncoupled system motion, we find that the energy is slightly shifted from the basic integral multiple of $\hbar\omega_0$ where ω_0 is the frequency of the individual oscillators in the uncoupled system.

The system we have been describing, the coupled oscillators, is a very elementary one, which we have been able to solve exactly with little difficulty. Suppose, however, that we have to find the solution in terms of independent, noninteracting systems oscillating with frequencies ω_0; then we must use perturbation theory. The zero-order Schroedinger wave equation for this system is

$$\frac{\partial^2\psi^0}{\partial x_1{}^2} + \frac{\partial^2\psi^0}{\partial x_2{}^2} + \frac{2m}{\hbar^2}[E^0 - \tfrac{1}{2}\omega_0{}^2(x_1{}^2 + x_2{}^2)]\psi^0 = 0 \qquad (11.13)$$

whose variables are separable, resulting in two equations, that for φ_1 being

$$\frac{\partial^2\varphi_1{}^0}{\partial x_1{}^2} + \frac{2m}{\hbar^2}[E_1{}^0 - \tfrac{1}{2}\omega_0{}^2 x_1{}^2]\varphi_1{}^0 = 0 \qquad (11.14)$$

Let $\alpha = \sqrt{m\omega_0/\hbar}$; then the zero-order functions are the one-dimensional harmonic oscillator wave functions

$$\psi^0 = H_n(\alpha x_1)e^{-\alpha^2 x_1{}^2/2} H_k(\alpha x_2)e^{-\alpha^2 x_2{}^2/2} \qquad (11.15)$$

where H_n is a normalized Hermite polynomial of the nth degree, and

$$E^0 = (n + k + 1)\hbar\omega_0 \qquad (11.16)$$

We wish to examine a degenerate system; the level with $E^0 = 2\hbar\omega_0$ is twofold degenerate since both $n = 1$, $k = 0$ and $n = 0$, $k = 1$ are states having this energy. For this level the elements of the secular determinant for the perturbation matrix Eq. (9.35) become

$$\begin{vmatrix} (V'_{10,10} - E'_i) & V'_{10,01} \\ V'_{01,10} & (V'_{01,01} - E'_i) \end{vmatrix} = 0 \qquad (11.17)$$

where the four matrix elements are

$$V'_{10,10} - E'_i = \int \{H_1(\alpha x_1)e^{-\alpha^2 x_1{}^2/2} H_0(\alpha x_2)e^{-\alpha^2 x_2{}^2/2} \Omega^2 x_1 x_2$$

$$\times H_1(\alpha x_1)e^{-\alpha^2 x_1{}^2/2} H_0(\alpha x_2)e^{-\alpha^2 x_2{}^2/2}\} dx_1 dx_2 - E'_i$$

$$= - E'_i \qquad (11.18)$$

since the integral is zero;

$$V'_{10,01} = \int \{H_1(\alpha x_1)e^{-\alpha^2 x_1{}^2/2} H_0(\alpha x_2)e^{-\alpha^2 x_2{}^2/2} \Omega^2 x_1 x_2$$

$$\times H_0(\alpha x_1)e^{-\alpha^2 x_1{}^2/2} H_1(\alpha x_2)e^{-\alpha^2 x_2{}^2/2}\} dx_1 dx_2$$

$$= \hbar\Omega^2/2\omega_0 \qquad (11.19)$$

$$V'_{01\ 10} = \hbar\Omega^2/2\omega_0 \qquad (11.20)$$

$$V'_{01,01} - E'_i = - E_i \qquad (11.21)$$

Finally, the energy of the perturbations is

$$E_t'^2 = \hbar^2\Omega^4/4\omega_0^2, \quad E'_t = \pm \hbar\Omega^2/2\omega_0 \tag{11.22}$$

which yields the same frequencies as Eqs. (11.4, 11.12).

Note that if we let

$$\psi_S = \frac{1}{\sqrt{2}}\{H_1(\alpha x_1)H_0(\alpha x_2) + H_0(\alpha x_1)H_1(\alpha x_2)\}\, e^{-\alpha^2(x_1{}^2+x_2{}^2)/2}$$

$$\psi_A = \frac{1}{\sqrt{2}}\{H_1(\alpha x_1)H_0(\alpha x_2) - H_0(\alpha x_1)H_1(\alpha x_2)\}\, e^{-\alpha^2(x_1{}^2+x_2{}^2)/2} \tag{11.23}$$

(where S and A refer to symmetry and antisymmetry, which will be taken up in the next chapter), the determinant Eq. (11.17) becomes diagonal,

$$\begin{vmatrix} (\hbar\Omega^2/2\omega_0) - E'_t & 0 \\ 0 & (-\hbar\Omega^2/2\omega_0) - E'_t \end{vmatrix} = 0 \tag{11.24}$$

showing that the wave functions Eq. (11.23) are the correct zero-order wave functions (i.e., the perturbation matrix is diagonal when they are used).

We can further emphasize that the above simple twofold degenerate problem is analogous with classical resonance by discussing it in terms of time-dependent perturbation theory. Assume that all but a_1 and a_2 are zero in Eq. (10.10) so that

$$i\hbar\dot{a}_1(t) = a_1(t)H'_{10,10} + a_2(t)H'_{10,01}$$
$$i\hbar\dot{a}_2(t) = a_1(t)H'_{01,10} + a_2(t)H'_{01,01} \tag{11.25}$$

since $E_{10} = E_{01}$ for this degenerate system. Let

$$b_+ = \frac{1}{\sqrt{2}}(a_1 + a_2); \quad b_- = \frac{1}{\sqrt{2}}(a_1 - a_2) \tag{11.26}$$

then we have

$$i\hbar\dot{b}_+ = b_+H'_{10,10} + b_+H'_{10,01}$$
$$i\hbar\dot{b}_- = b_-H'_{10,10} - b_-H'_{10,01} \tag{11.27}$$

in which we have used the fact that $H'_{01,01} = H'_{10,10}$ and $H'_{01,10} = H'_{10,01}$ from Eqs. (11.18)–(11.21).

Equations (11.27) have the solutions

$$\dot{b}_+ = A \exp\{-(i/\hbar)H'_{10,10}t - (i/\hbar)H'_{10,01}t\}$$

$$\dot{b}_- = B \exp\{-(i/\hbar)H'_{10,10}t + (i/\hbar)H'_{10,01}t\} \qquad (11.28)$$

When $B = A$, satisfying the initial condition that oscillator 1 is vibrating at maximum amplitude and oscillator 2 is at rest at $t = 0$, we have the normalized solutions for a_1 and a_2 of (11.25):

$$a_1 = \exp\{-(i/\hbar)H'_{10,10}t\} \cos(H'_{10,01}t/\hbar)$$

$$a_2 = \exp\{-(i/\hbar)H'_{10,10}t - i\pi/2\} \sin(H'_{10,01}t/\hbar) \qquad (11.29)$$

The probability of finding the system in the state ψ_{10} in which oscillator 1 is vibrating and oscillator 2 is still, is found from $a_1^*a_1$:

$$a_1^*a_1 = \cos^2(H'_{10,01}t/\hbar) = \cos^2(\Omega^2 t/2\omega_0) \qquad (11.30)$$

Similarly, the likelihood of the reverse situation, represented by ψ_{01}, is

$$a_2^*a_2 = \sin^2(\Omega^2 t/2\omega_0) \qquad (11.31)$$

The total wave function is given by

$$\psi = a_1(t)\psi_{10} + a_2(t)\psi_{01}$$

$$= \{\cos(\Omega^2 t/2\omega_0)H_1(\alpha x_1)H_0(\alpha x_2) - i\sin(\Omega^2 t/2\omega_0)H_0(\alpha x_1)H_1(\alpha x_2)\}$$

$$\times \frac{e^{-\alpha^2(x_1{}^2+x_2{}^2)}}{2e^{-i2\omega_0 t}} \qquad (11.32)$$

The reader will find it instructive to compare Eqs. (11.30–11.32) with that for the classical problem, Eq. (11.7).

The system may be said to resonate between the states ψ_{10} and ψ_{01} with a resonance energy $\pm \hbar\Omega^2/2\omega_0$ (Eq. 11.22). This energy must be assumed small compared to the zero-order unperturbed energy $2\hbar\omega_0$ in order for perturbation theory, upon which the theory of chemical resonance depends, to work.

11.3 CHANGE IN NOTATION

In order to simplify our work we adopt a much more convenient notation for wave functions and matrix elements, due to Dirac. Assuming ψ a wave function for a single particle with quantum numbers n, l, m, we will now write ψ as

$$\psi_{n,l,m} = |n, l, m\rangle, \quad \text{a ket vector}$$

$$\psi^*_{n,l,m} = \langle n, l, m|, \quad \text{a bra vector} \tag{11.33}$$

and

$$\int \psi^*_{n,l,m}\psi_{n,l,m}\,d\tau = \langle n, l, m|n, l, m\rangle$$

where the last is a help in remembering the names of these quantities if we think of it as a bra-ket or bracket. In these expressions n is the total quantum number, l the orbital momentum, and m the magnetic quantum number. If we are not interested in one of the quantum numbers, say m, we may drop it out of our label. If the total wave functions contain additional quantum numbers of interest, we may include them also. The matrix element or expectation value of an operator is written with the operator between the vectors, e.g. $\langle n,l,m|0|n,l,m\rangle$. For example, in the previous section our one-dimensional, two-oscillator wave function would be labeled $|n_1, n_2\rangle$. The matrix element given by Eq. (11.19) would then be expressed

$$V'_{10,01} \equiv \langle 10|\Omega^2 x_1 x_2|01\rangle \tag{11.34}$$

11.4 THE HELIUM ATOM

The normal, ground state helium atom has two electrons completely filling the ground state, and since only one nondegenerate state is possible, the wave functions were treated satisfactorily in Chapter 9 ignoring resonance. The first excited state of helium is degenerate, however, and in an excited helium atom resonance does occur in which either electron may spend some time in the higher energy level. The resonance energy that will be shown to result from this situation shifts the energy of the first excited P states and is observed in the shift of the frequency of the bright-line emitted radiation when the helium atom is de-excited and collapses to its ground state.

The zero-order wave functions for the helium atom, in which the interaction of the two electrons is ignored, are hydrogen-like wave functions. Because of the degeneracy between states of different n, l, and m there are eight states with the same energy possible to the unperturbed first excited energy level: if, say, electron number one is in the ground state $|100\rangle$, electron number two m.·v be in any one of the excited states $|200\rangle$, $|210\rangle$, $|211\rangle$, or $|21-1\rangle$, and conversely for electron number two in the ground state, the resulting four more states being physically indistinguishable from the first four. The total wave function would be written, for example, $|100, 200\rangle$ if the first electron was in the 100 state (also called 1s) and the second in the 200 (or 2s) state.

The interaction energy between the two electrons is treated by means of degenerate perturbation theory. The perturbation matrix elements for the energy of interaction between any two electron states are abbreviated by the letters J and K with subscripts s and p depending on whether both electrons are in an s state ($l = 0$) or one of them is in a p state ($l = 1$).

$$J_s = \langle 100, 200 \left| \frac{e^2}{r_{12}} \right| 100, 200\rangle \equiv \int \psi_{100}^{(1)*}\psi_{200}^{(2)*} \frac{e^2}{r_{12}} \psi_{100}^{(1)}\psi_{200}^{(2)} \, d\tau_1 d\tau_2$$

$$(11.35)$$

where (1) and (2) refer to each of the two electrons, $d\tau_1$ is the differential volume element $r_1^2 \sin\theta_1 \, dr_1 d\theta_1 d\varphi_1$ over the three degrees of freedom available to the first electron, and similarly for $d\tau_2$. $\langle 100, 200|$ and $|100, 200\rangle$ are the product of two hydrogen-like wave functions derived in Section 8.6 involving the Laguerre polynomials. J_s is also written for

$$\langle 200, 100 \left| \frac{e^2}{r_{12}} \right| 200, 100 \rangle$$

which has an identical numerical value.

$$K_s = \langle 100, 200 \left| \frac{e^2}{r_{12}} \right| 200, 100\rangle \equiv \int \psi_{100}^{(1)*}\psi_{200}^{(2)*} \frac{e^2}{r_{12}} \psi_{200}^{(1)}\psi_{100}^{(2)} \, d\tau_1 d\tau_2$$

$$(11.36)$$

or

$$\langle 200, 100 \left| \frac{e^2}{r_{12}} \right| 100, 200\rangle$$

which has the same numerical value. J_p, and also K_p, is written for six different matrix elements all having the same numerical result, one of which is

$$J_p = \langle 100, 211 \left| \frac{e^2}{r_{12}} \right| 100, 211\rangle$$

also,

$$K_p = \langle 100, 211 \left| \frac{e^2}{r_{12}} \right| 211, 100 \rangle \tag{11.37}$$

The other J_p and K_p matrix elements differ from these only in having the two electrons interchanged or in the orientation of the total angular momentum in space (that is, in having $m = 0$ or -1 instead of $+1$). The matrix elements other than these 16 elements are zero. For example, consider the matrix element $\langle 100, 200 | e^2/r_{12} | 100, 211 \rangle$: the $|211 \rangle$ wave function is an odd function as will be shown in Chapter 12, the other terms are all even, and thus, integrated over all space this matrix element must be zero. The J integrals are called *Coulomb integrals* since they give the potential energy contribution from the mutual repulsion of the two electrons. The K integrals are called *exchange integrals* since they result from the exchange of the two electrons in the wave function on the left as compared with the wave function on the right in the matrix element. Another name for the K integrals is *resonance integrals* since this energy contribution arises from the resonance of the two electrons between the states on the left and right of the matrix element.

By use of these matrix elements between the various hydrogen-like wave functions, the secular determinant Eq. (9.35) is written

$$\begin{bmatrix} (J_s - \Delta E) & K_s & 0 & 0 & 0 & 0 & 0 & 0 \\ K_s & (J_s - \Delta E) & 0 & 0 & 0 & 0 & 0 & 0 \\ 0 & 0 & (J_p - \Delta E) & K_p & 0 & 0 & 0 & 0 \\ 0 & 0 & K_p & (J_p - \Delta E) & 0 & 0 & 0 & 0 \\ 0 & 0 & 0 & 0 & (J_p - \Delta E) & K_p & 0 & 0 \\ 0 & 0 & 0 & 0 & K_p & (J_p - \Delta E) & 0 & 0 \\ 0 & 0 & 0 & 0 & 0 & 0 & (J_p - \Delta E) & K_p \\ 0 & 0 & 0 & 0 & 0 & 0 & K_p & (J_p - \Delta E) \end{bmatrix} = 0$$

$$\tag{11.38}$$

where ΔE is the small energy shift due to the perturbation.

Equation (11.38) may be rewritten

$$[(\Delta E - J_s)^2 - K_s^2][(\Delta E - J_p)^2 - K_p^2]^3 = 0 \tag{11.38a}$$

since all the J's and K's with the same subscript are equal. The solutions of (11.38a) are

$$\Delta E = J_s + K_s, \quad J_s - K_s, \quad J_p + K_p, \quad J_p - K_p \qquad (11.39)$$

with the last two terms still giving a triple degeneracy since the degeneracy in m has not been broken up. Since the energy does not depend on m, for convenience we may drop the magnetic quantum number from the bra and ket vectors for the rest of this section. The splitting up of the degeneracy between $|10, 20\rangle$ and $|10, 21\rangle$ (i.e., the $2s$ and $2p$ states, see Figure 11.3), corresponding to the difference between J_s and J_p was said in Chapter 8 to be due to the greater penetration of lower l electrons within the electron cloud. The further splitting of these states, however, results from the resonance energy expressed by the exchange integrals. We have also discussed previously the raising of the ground state due to the mutual shielding of the positive nucleus by the electrons. Figure 11.3 illustrates the effect of these perturbation calculations.

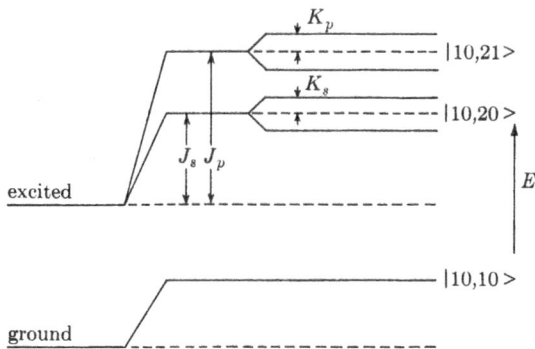

Figure 11.3 *Illustration of the effect of the interaction energy of the two electrons on the lowest two unperturbed levels for the electrons in the helium atom. The two lines on the left represent the ground state energy and first excited state energy if the perturbation caused by the electron repulsion is neglected. Both levels are raised by the electron shielding of the nuclear electrostatic potential, but the excited level is split into several levels as a result of differences in shielding between s ($l = 0$) and p ($l = 1$) electrons and quantum mechanical resonance.*

Thus, we have found that the eightfold degenerate first excited level of the helium atom splits into four levels, of which the lower two are entirely nondegenerate due to the addition or subtraction of the resonance energy. If we choose new wave functions

$$\psi_S = \frac{1}{\sqrt{2}}(|100, 200\rangle + |200, 100\rangle)$$

$$\psi_A = \frac{1}{\sqrt{2}}(|100, 200\rangle - |200, 100\rangle) \qquad (11.40)$$

then K_s in Eq. (11.36) becomes

$$
\begin{aligned}
K_s &= \int \psi_S{}^* \frac{e^2}{r_{12}} \psi_A \\
&= \frac{1}{2}\Bigg(\langle 100, 200 \Big| \frac{e^2}{r_{12}} \Big| 100, 200\rangle - \langle 100, 200 \Big| \frac{e^2}{r_{12}} \Big| 200, 100\rangle \\
&\quad + \langle 200, 100 \Big| \frac{e^2}{r_{12}} \Big| 100, 200\rangle - \langle 200, 100 \Big| \frac{e^2}{r_{12}} \Big| 200, 100\rangle \Bigg) = 0
\end{aligned}
$$

$$(11.41)$$

Similarly, K_p also vanishes when the wave functions (11.40) are used. Similarly, the nonzero J integrals are found to be symmetric and antisymmetric.

$$J_{ss} = \langle 100, 200 \Big| \frac{e^2}{r_{12}} \Big| 100, 200\rangle + \langle 100, 200 \Big| \frac{e^2}{r_{12}} \Big| 200, 100\rangle$$

$$J_{sa} = \langle 100, 200 \Big| \frac{e^2}{r_{12}} \Big| 100, 200\rangle - \langle 100, 200 \Big| \frac{e^2}{r_{12}} \Big| 200, 100\rangle \qquad (11.42)$$

ψ_S is called a *symmetric* wave function since it does not change sign if the two electrons are interchanged. ψ_A, however, does change sign and is thus an *antisymmetric* wave function. ψ_A and ψ_S are seen to be the correct first-order wave functions since the off-diagonal elements K_s, K_p vanish if these wave functions are used. Whereas before inclusion of the interaction of the electrons only one energy level was predicted, now four closely spaced energy levels are predicted. The lowest level is due to the antisymmetric 2s state, the next to the symmetric $2s$ state, the third to the antisymmetric degenerate $2p$ level, and the highest to the symmetric $2p$ level.

In general, then, whenever a system comprised of two similar members may have two exactly similar states, there will be a lowering and raising (i.e. a splitting) of the energy of these states compared to the energy calculated regarding just one member of the system. The depression of the energy level is due to resonance between the two possible degenerate states.

11.5 THE HYDROGEN MOLECULE (HEITLER–LONDON TREATMENT)

In the diatomic hydrogen molecule the two electrons can alternate in orbiting around one proton or the other; thus the hydrogen molecule ground state furnishes us with another simple example of resonance. In general, the hydrogen–hydrogen bond is composed of two types of states, one in which both electrons surround one proton forming an ionic bond and the other type in which each proton has an electron close to it. This latter type of state, corresponding to a perturbation of the wave functions for infinitely separated hydrogen atoms, is the one we will assume to predominate. Figure 11.4 illustrates the change in electron potential and consequent change in wave functions as the two atoms are brought close together. A continuous wave function and derivative cannot be realized by joining the two infinite-separation wave functions when the two atoms are joined. Only by lowering the energy eigenvalue of the system (or raising it to a new first excited level) can a smooth fit for the total wave function be made.

Thus we consider the two atoms well separated as compared to the distance of the electrons from the closest protons. The initial wave function will then consist of the product of the ordinary ground state hydrogen wave functions for each atom, $\psi_A(\mathbf{r}_1)\psi_B(\mathbf{r}_2)$, and will clearly yield resonance energy. Let r_{1A} represent the separation of electron 1 from proton A, and similarly r_{2B} for electron 2 and proton B. The complete wave equation is then

$$\left[-\frac{\hbar^2}{2m}(\nabla_1^2 + \nabla_2^2) - \frac{e^2}{r_{1A}} - \frac{e^2}{r_{2B}} - \frac{e^2}{r_{1B}} - \frac{e^2}{r_{2A}} + \frac{e^2}{r_{12}} + \frac{e^2}{r_{AB}} \right]\psi = E\psi$$

The unperturbed function is $\psi = \psi_A(\mathbf{r}_1)\psi_B(\mathbf{r}_2)$ where $\psi_A(\mathbf{r}_1)$ is a solution to

$$\left(-\frac{\hbar^2}{2m}\nabla_1^2 - \frac{e^2}{r_{1A}}\right)\psi_A(\mathbf{r}_1) = E_1\psi_A(\mathbf{r}_1)$$

$\psi_A(\mathbf{r}_1)$ is the associated Laguerre and spherical harmonic function of Chapter

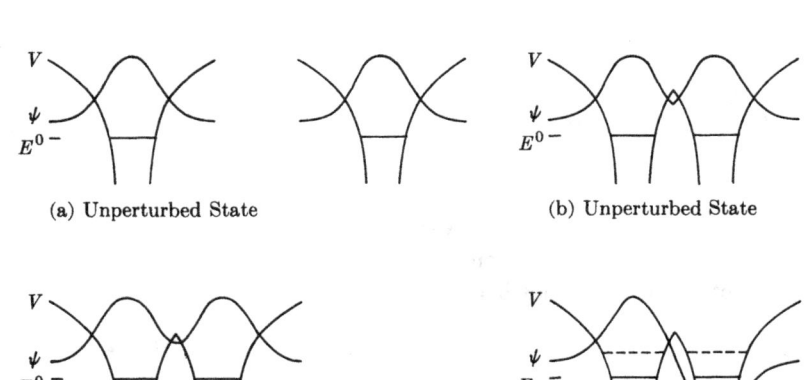

(a) Unperturbed State   (b) Unperturbed State

(c) Perturbed Ground State   (d) Perturbed Excited State

Figure 11.4 *Illustration of the behavior of the potential energy, the energy levels, and the complete two-electron wave functions for two hydrogen atoms (a) infinitely separated (b) brought together with the separated wave functions drawn (unchanged from a), (c) brought together and showing the corrected, smoothly joining wave functions, (d) brought together and showing the correct wave function for the first excited state. The dashed lines in (c) and (d) are for the correct energy levels appropriate to the wave functions shown in (c) and (d). The straight solid horizontal lines on all four figures shows the ground energy level of the infinitely separated atoms, for comparison.*

8, and similarly for $\psi_B(\mathbf{r}_2)$. Let

$$H' = -\frac{e^2}{r_{1B}} - \frac{e^2}{r_{2A}} + \frac{e^2}{r_{12}} + \frac{e^2}{r_{AB}}$$

Then the degenerate perturbation integrals are

$$J = \int\int \psi_A{}^*(\mathbf{r}_1)\psi_B{}^*(\mathbf{r}_2) \; H' \; \psi_A(\mathbf{r}_1)\psi_B(\mathbf{r}_2) \; d\tau_1 d\tau_2$$

$$K = \int\int \psi_A{}^*(\mathbf{r}_1)\psi_B{}^*(\mathbf{r}_2) \; H' \; \psi_A(\mathbf{r}_2)\psi_B(\mathbf{r}_1) \; d\tau_1 d\tau_2$$

$$+ \int\int \psi_A{}^*(\mathbf{r}_1)\psi_B{}^*(\mathbf{r}_2) \; H^0 \; \psi_A(\mathbf{r}_2)\psi_B(\mathbf{r}_1) \; d\tau_1 d\tau_2 \tag{11.43}$$

where $d\tau_1$ and $d\tau_2$ are the differential spherical volume elements

$$d\tau_1 = r_1{}^2 \sin\theta_1 \, dr_1 d\theta_1 d\varphi_1$$

The second term in the exchange integral K, which we will label K_0, must be included since our initial wave functions are not orthogonal, i.e., $I_{jk'}$ of Eq. (9.33) is not here given by $I_{jk'} = \delta_{jk'}$. Since

$$H^0 \, \psi_A(\mathbf{r}_2)\psi_B(\mathbf{r}_1) = E^0 \, \psi_A(\mathbf{r}_2)\psi_B(\mathbf{r}_1) = 2E_1 \, \psi_A(\mathbf{r}_2)\psi_B(\mathbf{r}_1)$$

we have

$$K_0 = 2E_1 \int\int \psi_A{}^*(\mathbf{r}_1)\psi_B{}^*(\mathbf{r}_2) \, \psi_A(\mathbf{r}_2)\psi_B(\mathbf{r}_1) \; d\tau_1 d\tau_2$$

The contribution of e^2/r_{AB} to J is simply e^2/r_{AB}, since the wave functions are normalized and e^2/r_{AB} may be taken outside the integration over the electron coordinates. Similarly, the contribution of the e^2/r_{AB} term to K is simply $[(e^2/r_{AB})/2E_1]K_0$. The contribution of (e^2/r_{1B}) to J would be

$$\int \psi_A{}^*(\mathbf{r}_1)\left(\frac{-e^2}{r_{1B}}\right)\psi_A(\mathbf{r}_1) \, d\tau_1$$

since $\psi_B(\mathbf{r}_2)$ is normalized and r_{1B} is not a function of the position of the second electron. The contribution of (e^2/r_{1B}) to K would be similar.

We thus obtain the secular determinant

$$\begin{vmatrix} (J - \Delta E) & K \\ K & (J - \Delta E) \end{vmatrix} = 0 \tag{11.44}$$

from which the shifts in energy level caused by the resonance energy is found to be

$$\Delta E = J + K, \quad J - K$$

The Coulomb effects between all the charges roughly cancel out, leaving the exchange effects which cause chemical binding, as illustrated in Figure 11.5.

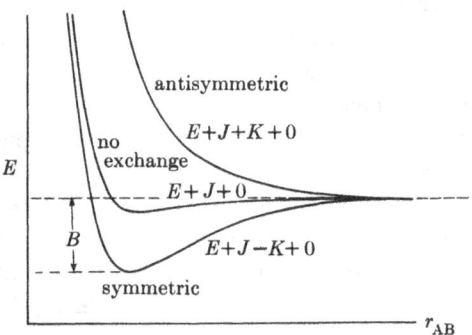

Figure 11.5 *Illustration of effect of exchange or resonance energy on the binding energy, B, of the hydrogen molecule. The middle curve includes all effects J + 0 other than the resonance of the two electrons where 0 represents all interactions not specifically calculated in the text.*

It is found that the symmetric wave function

$$\psi_S = \frac{1}{\sqrt{2 + 2K_0}}[\psi_A(\mathbf{r}_1)\psi_B(\mathbf{r}_2) + \psi_A(\mathbf{r}_2)\psi_B(\mathbf{r}_1)] \qquad (11.45)$$

is the correct zero-order wave function which with its antisymmetric twin causes the perturbation matrix to be diagonal. Equation (11.45) is the symmetric wave function whose eigenvalue is approximately $E_0 - K$, the lower of the two energy eigenvalues including resonance. This is in contrast to the case of helium where both electrons are in the same atom, and the antisymmetric wave function was found to be the lower of the two energy eigenvalues including resonance. The attractive bonding force of the hydrogen molecule has been shown to be due to the possibility of two electrons being in a symmetric state (i.e., being interchangeable in the wave function representation). We will see in the next chapter that no more than two electrons can be in one symmetric state at the same time, so that the degeneracy in this case can be only twofold.

The symmetric wave function Eq. (11.45) yields a probability of finding both electrons between the two protons larger than the probability of this event for the two atoms placed a distance r_{AB} apart and no resonance effects taken into account. That is, $|\psi_A(\mathbf{r}_1)\psi_B(\mathbf{r}_2)|^2$ gives a probability of this event smaller than that given by the absolute square of the symmetric wave function Eq. (11.45) (see Figure 11.5). The antisymmetric wave function, on the other hand, gives a smaller electron probability density in this region, compared to the nonresonance value. Thus, despite the mutual repulsion of the two electrons, the attraction of the two protons for the (symmetric state) negative cloud causes the hydrogen molecule to be a stable configuration. The mutual repulsion of the less screened protons in the antisymmetric electronic state results in an unstable configuration. ΔE, J, and K are functions of internuclear distance as shown in Figure 11.5.

Problem 11.1 The interactions of two particles in the same ground state in a one-dimensional square well of width a cm is a small square well-type potential which is V_0 ev deep and b cm wide. That is, $V' = 0$ unless $x_2 - x_1 \leqslant b$, where the x's are particle coordinates, so that the integral over x_2 extends from $x_1 - b$ to $x_1 + b$ and the subsequent integral over x_1 from 0 to a. How does their interaction affect their energy? If the particles repel one another with a potential $+V_0$, how does this affect the symmetry of the ground state?

Problem 11.2 If two dipoles with moments $e\mathbf{l}_1$ and $e\mathbf{l}_2$ (e being the electronic charge and $\mathbf{l}_1$ and $\mathbf{l}_2$ being the charge separation in the two dipoles) are a large distance r apart, their interaction energy is

$$V = \frac{e^2}{r^3}[\mathbf{l}_1 \cdot \mathbf{l}_2 - 3(\mathbf{l}_1 \cdot \hat{\mathbf{r}})(\mathbf{l}_2 \cdot \hat{\mathbf{r}})]$$

where $\hat{\mathbf{r}}$ is a unit vector in the $\mathbf{r}$ direction. What is the interaction energy between an excited hydrogen atom ($n = 2$, $l = 1$, $m = 0$) and an unexcited hydrogen atom r cm away? Assume the polar axis of the excited atom to be parallel to $\mathbf{r}$. Note that if two oscillating dipoles are in phase with one another they will attract one another, whereas if out of phase they will repel one another.

11.6 ELECTROSTATIC INTERACTION BETWEEN TWO ELECTRONS IN GENERAL

The dominant force on atomic electrons is the attraction of the multiply-charged positive nucleus. Because the forces between the various atomic electrons is less than this, we have neglected the presence of other electrons in zero-order approximation. Hence, our zero-order approximate wave-functions have been hydrogen wave functions—products of Laguerre and associated spherical harmonic functions. The more complicated interaction with the other electrons in an atom or in neighboring atoms have then been treated as a perturbation. The electronic interactions, as we have seen in Sections 11.4 and 11.5, are crucial to a prediction of atomic energy levels and to understanding chemical binding. Thus, the interaction between two electrons in hydrogen electronic states is of recurring interest, and it is very useful to review the results of Sections 11.4 and 11.5 and see what general conclusions or selection rules we can deduce whenever the interaction of two electrons must be considered.

Using perturbation theory we obtain matrix elements, that is, integrals of the type

$$\int \psi^a(1)\psi^b(2) \frac{1}{r_{12}} \psi^c(1)\psi^d(2)\, d\tau_1 d\tau_2$$

If $a = c$ and $b = d$, the integrals are the Coulomb integrals, whereas if $a = d$ and $b = c$, the integrals are the exchange or resonance integrals (see Eqs. 11.36–11.38). $1/r_{12}$ may be expanded in terms of the spherical harmonics since these form a complete orthonormal set. The expansion is

$$\frac{1}{r_{12}} = \sum_{k=0}^{\infty} \frac{r_<^k}{r_>^{k+1}} \frac{4\pi}{2k+1} \sum_{m=-k}^{+k} Y_{k,m}(\theta_1,\varphi_1) Y_{k,-m}(\theta_2,\varphi_2)$$

$$= \sum_{k=0}^{\infty} \frac{r_<^k}{r_>^{k+1}} \frac{4\pi}{2k+1} \sum_{m=k}^{k} \Theta_{km}(\theta_1)\Phi_m(\varphi_1)\, \Theta_{km}(\theta_2)\Phi_{-m}(\varphi_2) \qquad (11.46)$$

where $r_<$ denotes the radial coordinate of the inner electron (the one closer to the nucleus) and $r_>$ denotes the radius of the outer electron.

The one-electron wave function is the hydrogen wave function which for electron number one, for example, in the ath state, is written in terms of Laguerre and spherical harmonic functions,

$$\psi^a(1) = R_{n_a l_a}(r_1)\Theta_{l_a m_a}(\theta_1)\Phi_{m_a}(\varphi_1)$$

The integral over φ_1 is trivial:

$$\int_0^{2\pi} \Phi_{m_a}^*(\varphi_1)\Phi_{m_c}(\varphi_1)\Phi_m(\varphi_1) \, d\varphi_1$$

$$= \frac{1}{(2\pi)^{3/2}} \int_0^{2\pi} e^{-im_a\varphi_1} e^{im_c} e^{im\varphi} \, d\varphi_1$$

$$= \frac{1}{\sqrt{2\pi}} \delta_{m_a, m_c+m} \tag{11.47}$$

Similarly the integral over φ_2 yields

$$\frac{1}{\sqrt{2\pi}} \delta_{m_b, m_d-m}$$

Hence,

$$m = m_a - m_c = m_d - m_b$$

$$m_a + m_b = m_c + m_d \tag{11.48}$$

Only the term for which (11.48) is satisfied is nonzero in the sum over m in (11.46), and thus the z-axis projection of the two electrons' total angular momentum is the same in both the ab and cd states. The θ integrals, which are tabulated in Chapter 6 of Condon and Shortley (see Bibliography of Chap. 12), are

$$\int_0^{\pi} \Theta_{l_a m_a}(\theta_1) \, \Theta_{l_c m_c}(\theta_1) \, \Theta_{k, m_a-m_c}(\theta_1) \sin\theta_1 \, d\theta_1 \equiv C^k(l_a m_a, l_c m_c) \tag{11.49}$$

where k refers to the number of one of the terms in the power series expansion in $r_<$ and $r_>$ in (11.46), and $\Theta_{l,m}(\theta) \propto (\cos\theta)^{l-m}$ plus terms of lower order whose exponent is less by an even integer; hence the integrand in Eq. (11.49) is proportional to $(\cos\theta)^{l_a+l_c+k-2m_a}$. If $l_a + l_c + k$ is odd, the integrand is odd and the integral is zero; hence no mixture of states for which $l_a + l_c + k$

is odd results. Besides the evenness of $l_a + l_c + k$ we obtain the following selection rules:

$$k \geqslant |l_a - l_c|, \quad k \geqslant |m_a - m_c|, \quad \text{and} \quad k < l_a + l_c \qquad (11.50)$$

11.7 RELATIVE SIGNIFICANCE OF THE SEVERAL ATOMIC INTERACTIONS

In discussing various atomic examples of perturbation theory in Sections 9.2, 9.5, 9.6, 11.4, 11.5, 11.6 we always neglected the electronic interaction and the spin–orbit interaction on the electrons in zero-order approximation. Then in first-order approximation we used the hydrogen wave functions to estimate the change in energy levels and wave functions from either of these two perturbing effects. It is important to compare the magnitude of these perturbations with each other in order to know which perturbation should be used first in calculating energy levels and wave functions in higher order approximations. It is also important to compare the perturbations with the nuclear Coulombic potential for an indication of the reliability of the perturbation approximation.

We consider the magnitudes of the energy contributions of various atomic electron interactions as displayed in Table 11.1. Note that in light

Table 11.1 *Magnitudes of the energies (in ev) of the various interactions involving atomic electrons.*

Interaction effects	Light atoms	Valence electrons of heavy atoms	K or 1st Shell electrons of heavy atoms
Nuclear potential	$\sim$100–1000	$\sim$2–10	$\sim$20,000–100,000
Electrostatic repulsion of electrons	$\sim$1	$\sim$1	$\sim$100
Spin–orbit coupling	$\sim$10^{-4}–10^{-3}	$\sim$0.1–1.0	$\sim$3000

atoms ($Z \sim$ 2–10) the electrostatic interaction dominates the spin–orbit coupling. In general, the mutual repulsion of atomic electrons separates levels of different symmetry. This effect is called Russell–Saunders coupling; in light atoms Russell–Saunders coupling is of primary importance. For the inner electrons of heavy atoms, on the other hand, the **L · S** or spin–orbit coupling is predominant. (J commutes with the Hamiltonian including the **L · S** term whereas L and S do not. Therefore J is a constant of the motion. The large size of the **L · S** term causes states of different J to be widely separated.) In this case neither the magnetic spin quantum numbers nor the magnetic orbital momentum quantum number l is unique for the widely separated levels of different total angular momentum, $j = |{\bf l} + {\bf s}|$. Hence the $|j, m \rangle$ states having total angular momentum **j** and $j_z \equiv m$, the magnetic quantum number, are the important states for these electrons. The weaker electrostatic interaction slightly separates levels of different symmetry as a weak perturbation. Thus the total angular momentum has a primary effect on the energy as a result of the **L · S** interaction. This situation is called j–j coupling, and in inner electrons of heavy atoms is the predominant effect.

In both R–S and j–j coupling the higher-order perturbation calculation is carried out in the representation in which the predominant perturbation is diagonal. Unfortunately neither Russell–Saunders nor j–j coupling is appropriate to the intermediate case of the outer electrons in heavy atoms. Successive perturbation treatment is not possible, so that both perturbations must be handled simultaneously, which is much more messy and difficult. Similar changes in the relative importance of these two physical effects occur in nuclear physics, as one goes from one end of the periodic table to the other.

BIBLIOGRAPHY

Pauling, L., and E. B. Wilson, cited in Chapter 7.

SPIN, SYMMETRY,

PARITY, AND VECTOR ADDITION

12

UP TO THIS POINT we have concentrated on the energy and angular momentum properties of multiple body systems. Other mathematical and physical properties of the particles and their wave functions exist, however, which yield additional information about the system. These other properties, which are revealed by appropriate operators, are also frequently constants of the motion (they do not change with time) and it is useful to characterize states in terms of the eigenvalues of their operators. (*i*) The intrinsic angular momentum or *spin* of a particle or system of particles, as opposed to the orbital angular momentum with which we have been concerned, is one such property. The spin being a separate variable of the system requires a special eigenfunction which we will soon discuss, in Section 12.2. (*ii*) The spin eigenfunction of a system of two particles combines with the spatial eigenfunctions to form wave functions which either do or do not change sign if the particles are interchanged. This exchange property of eigenfunctions (either total eigenfunctions or one of the component product eigenfunctions, for example spin or space functions) is called *symmetry*. It is an important and very useful property since the symmetry of a total eigenfunction for a particle is *always* the same depending on what kind of a particle it is. (*iii*) If the coordinate frame in which a system is represented is reversed so that a polar vector in one frame has an opposite sign in the other frame ($\mathbf{r} \rightarrow -\mathbf{r}$), the wave function may or may not change sign. This behavior of wave functions (total or spatial) under coordinate reversal is called *parity* and it too is a very interesting and useful property of the wave function of a system. Both symmetry and parity conditions restrict the kind and number of possible eigenfunctions a system may have and the transitions which the system may undergo; herein lies their greatest importance and utility.

In discussing spin and other topics of this chapter, it will be particularly useful to be able to represent various operators in matrix notation. We may explain matrix notation in the following way. Let $\varphi_1, \varphi_2 \ldots$ form a complete orthonormal set of wave functions, say the eigenfunctions of some operator A. As we already explained, any arbitrary wave function ψ can then be represented as a series

$$\psi = \sum a_n \varphi_n, \text{ where } a_n = \int d\tau \, \varphi_n^* \psi.$$

Indeed we can specify the function by giving only the set of coefficients ($a_1, a_2 \ldots$), since the series can always be reconstructed. There is a one-one correspondence, in other words, between wave functions ψ and ordered sets of numbers (i.e., between a wave function and a vector $a_1, a_2 \ldots$, or $\{a_n\}$ for short). One can also set up a correspondence between operators and doubly ordered sets of numbers (i.e., between operators and matrices) as follows:

$$H_{mn} = \int d\tau \, \varphi_m^* H \varphi_n$$

Now let $x = H\psi$ and compute the coefficients $\{b_n\}$ of the new function x:

$$b_m = \int d\tau \, \varphi_m^* x$$

$$= \int d\tau \, \varphi_m^* H \psi$$

$$= \sum_n \left[\int d\tau \, \varphi_m^* H \varphi_n \right] a_n = \sum_n H_{mn} a_n,$$

which is the familiar rule for the multiplication of a vector by a matrix. We can think of a wave function ψ and an operator H, therefore, as a vector $\{a_n\}$ and a matrix H_{mn}; and if we do, the result of letting the operator act on the wave function will be the same as the result of multiplying the vector by the matrix. (Note, by the way, that the vectors and matrices are generally infinite–dimensional.)

The next thing to show is that the product of two operators corresponds in the proper way to the product of the two matrices. Let $C = AB$, where A, B, and C are operators. Then

$$C_{mn} = \int d\tau \, \varphi_m{}^* C \varphi_n$$

$$= \int d\tau \, \varphi_m{}^* A B \varphi_n$$

Now the series for $B\varphi_n$ is

$$B\varphi_n = \sum_p \left(\int d\tau \, \varphi_p{}^* B \varphi_n \right) \varphi_p$$

$$= \sum_p B_{pn} \varphi_p$$

Hence

$$C_{mn} = \int d\tau \, \varphi_m{}^* A \left(\sum_p B_{pn} \varphi_p \right)$$

$$= \sum_p B_{pn} \int d\tau \, \varphi_m{}^* A \varphi_p$$

$$= \sum_p A_{mp} B_{pn}$$

But this is simply the product of the two matrices which is what we set out to prove.

It should be emphasized that the matrix elements depend on the choice of the orthonormal set φ_1, φ_2 . . . In other words, there is more than one matrix representation. If the φ_n's are the eigenfunctions of some operator A, one speaks of the A representation. For example, if they are the simultaneous eigenfunctions of the operators H, L^2, and L_z of Chapter 8, we have the so-called energy-angular momentum representation. An operator is said to be *diagonal* in a certain representation if its matrix elements vanish when the indices are different.

$$A_{mn} = 0 \text{ whenever } m \neq n.$$

(*Exercise:* Show that the operator A is diagonal in the A representation, that is the representation in which $A\varphi_n = a_n\varphi_n$.) It is very often important to *find* a representation in which a given operator (usually the Hamiltonian) is diagonal, and the process of doing so is called *diagonalization*. This, in fact, is what we were doing in the second-order perturbation theory; we were finding an orthonormal set of wave functions (eigenfunctions of H_0) in which the operator V was diagonal.

We might review the material of previous chapters with respect to their matrix representation. In this regard, degenerate perturbation theory provides a good illustrative introduction to the significance of and requisite conditions for diagonality of an operator matrix. For example, when $H^0\psi_{tk}{}^0 = E_t{}^0\psi_{tk}{}^0$, we say that the Hamiltonian or energy operator is diagonal in this matrix representation, for H^0 operating on any of these ψ^0's always gives us back the same wave function. If the ψ^0's are orthogonal, multiplying by $\psi_{1k'}{}^{0*}$ on the left and integrating over all space would yield

$$\langle \psi_{tk'}{}^{0*} H^0 \psi_{tk}{}^0 \rangle = E_t{}^0 \delta_{k'k} \tag{12.1}$$

The subscripts k and k' may have any values from 1 to n for an n-fold degenerate system. These n^2 numbers may be ordered in the form of a matrix where the first index refers to the row and the second index refers to the column. Thus Eq. (12.1) is written

$$
\begin{bmatrix}
\langle \psi_{t1}{}^{0*}H^0\psi_{t1}{}^0 \rangle & \langle \psi_{t1}{}^{0*}H^0\psi_{t2}{}^0 \rangle & \cdots & \langle \psi_{t1}{}^{0*}H^0\psi_{tn}{}^0 \rangle \\
\langle \psi_{t2}{}^0 H^0\psi_{t1} \rangle & \langle \psi_{t2}{}^{0*}H^0\psi_{t2} \rangle & \cdots & \langle \psi_{t2}{}^{0*}H^0\psi_{tn}{}^0 \rangle \\
\cdot & \cdot & \cdot & \\
\cdot & \cdot & \cdot & \\
\cdot & \cdot & \cdot & \\
\langle \psi_{tn}{}^0 H^0\psi_{t1}{}^0 \rangle & \langle \psi_{tn}{}^{0*}H^0\psi_{t2}{}^0 \rangle & \cdots & \langle \psi_{tn}{}^{0*}H^0\psi_{tn}{}^0 \rangle
\end{bmatrix}
= E_t{}^0
\begin{bmatrix}
1 & 0 & 0 & \cdots & 0 \\
0 & 1 & 0 & \cdots & 0 \\
0 & 0 & 1 & \cdots & 0 \\
\cdot & \cdot & \cdot & & \cdot \\
\cdot & \cdot & \cdot & & \cdot \\
\cdot & \cdot & \cdot & & \cdot \\
0 & 0 & 0 & \cdots & 1
\end{bmatrix}
$$

$$\tag{12.2}$$

If an additional term V' is added to the Hamiltonian operator, $\langle \psi_{ik'}{}^{0*} H \psi_{ik}{}^0 \rangle$ is no longer necessarily zero if $k \neq k'$, as we know. Hence we obtain a nondiagonal matrix, because V' operating on $\psi_{ik}{}^0$ yields a wave function different from $\psi_{ik}{}^0$. If

$$\int \psi_{ik'}{}^{0*} V' \psi_{ik}{}^0 \, d\tau \neq V' \delta_{k'k}$$

it is because $V' \psi_{ik}{}^0$ yields some $\psi_{ik'}{}^0$ (plus some other ψ^0's, in general), for $V' \psi_{ik}{}^0$ has to yield some off-diagonal ψ's in order for this equation to be true since $\int \psi_{ik'}{}^{0*} \psi_{ij}{}^0 \, d\tau = 0$ unless $k' = j$. In several of the simple examples treated in Chapter 11, we discovered other representations in which the energy was again diagonal; that is, we *diagonalized* the Hamiltonian operator.

12.1 MATRICES OF ANGULAR MOMENTUM

Besides the Hamiltonian or energy operator, another vital operator representing a physical observable is the square of the total angular momentum operator or of operators representing any of its components. While one almost always chooses a representation in which the energy is diagonal, in accordance with the central importance of the energy of a physical state, a given component of the angular momentum may not be diagonal in the representation chosen for a given system. Indeed, only one component of the angular momentum may be diagonal in any one representation, the other two components being necessarily nondiagonal. Thus L_x and L_y are nondiagonal in a representation in which L_z is diagonal. Let us for the moment confine our attention just to the orbital momentum.

Note that if L_x, for example, operates on a wave function such as $R(r) Y_l{}^m$ which is represented in spherical coordinates with polar axis along the z axis, new wave functions result (cf. Section 8.1, 2). L_z operating on these particular spherical harmonic wave functions yields the original wave function, and therefore $R(r) Y_l{}^m$ is an eigenfunction of L_z but not of L_x,

$$L_z R(r) Y_l{}^m(\theta, \varphi) = m\hbar R(r) Y_l{}^m(\theta, \varphi) \tag{12.3}$$

whereas

$$L_x R_{n,l}(r) Y_l^m(\theta, \varphi)$$

$$= \frac{\hbar}{i} \left\{ - \sin\varphi \frac{\partial}{\partial\theta} - \cot\theta\cos\varphi \frac{\partial}{\partial\varphi} \right\} R_{n,l}(r) \frac{(-1)^{(l+|m|)/2}}{\sqrt{4\pi}}$$

$$\times \sqrt{\frac{(2l+1)(l-|m|)!}{(l+|m|)!}} \, \Theta_l^{|m|}(\cos\theta) e^{im\varphi}$$

$$= \frac{\hbar}{2} \{ \sqrt{(l-m)(l+m+1)} R_{n\,l}(r) Y_l^{m+1}(\theta, \varphi)$$

$$+ \sqrt{(l-m+1)(l+m)} R_{n,l}(r) Y_l^{m-1}(\theta, \varphi) \} \tag{12.4}$$

where the Y_l^m have been written explicitly once as a product of a function of θ and $e^{im\varphi}$ to illustrate the functions the derivatives operate on. (The Y_l^m have been suitably normalized by the complicated function of l and m occurring in (12.4).) Hence use of the operator L_x in this representation results in a linear combination of new wave functions.

We will now put the angular momentum operator into a matrix representation. (It will be finite-dimensional representation, since we are going to consider only one value of l at a time. There will be one dimension for each value of m, and therefore $2l+1$ in all.) An eigenfunction of this state may be written as a linear combination of spherical harmonics having allowed m (i.e., $-l \leqslant m \leqslant l$):

$$\psi = c_l Y_l^l(\theta, \varphi) + c_{l-1} Y_l^{l-1}(\theta, \varphi) + \ldots + c_{-l} Y_l^{-l}(\theta, \varphi)$$

where the c's are constant coefficients giving the contribution of each particular Y_l^m to the eigenfunction. A different ψ would have different c's since it would be made up of a different combination of the $(2l+1) Y_l^m$'s. In short, if each Y_l^m is compared to a unit vector parallel to the x, y, or z axis in coordinate space, each constant coefficient acts as a component of any given eigenfunction in the same way as x, y, or z is a component of a vector. These component c's thus act like coordinates in a $(2l+1)$-dimension vector space, and ψ could be represented as a column vector

$$\psi = \begin{bmatrix} c_l \\ c_{l-1} \\ . \\ . \\ . \\ c_{-l} \end{bmatrix} \tag{12.5}$$

The operator L_z operating on Y_l^m must yield the same Y_l^m times a constant $(m\hbar)$, depending on the component of ψ it operates on. Hence in a representation with the z axis parallel to the polar axis, the operator L_z would be represented by a diagonal matrix

$$L_z = \hbar \begin{bmatrix} l & 0 & 0 & \ldots & 0 \\ 0 & (l-1) & 0 & \ldots & 0 \\ 0 & 0 & (l-2) & \ldots & 0 \\ . & . & . & & . \\ . & . & . & & . \\ . & . & . & & . \\ 0 & 0 & 0 & \ldots & -l \end{bmatrix} \tag{12.6}$$

while the operator L_x would have to be represented by a nondiagonal matrix since this operator introduces a different combination of Y_l^m's, Eq. (12.4):

$$L_x = \frac{\hbar}{2} \begin{bmatrix} 0 & a_l & 0 & 0 & \ldots & 0 & 0 \\ a_l & 0 & a_{l-1} & 0 & \ldots & 0 & 0 \\ 0 & a_{l-1} & 0 & a_{l-2} & \ldots & 0 & 0 \\ 0 & 0 & a_{l-2} & 0 & \ldots & 0 & 0 \\ . & . & . & . & & . & . \\ . & . & . & . & & . & . \\ . & . & . & . & & . & . \\ 0 & 0 & 0 & 0 & \ldots & 0 & a_{-l+1} \\ 0 & 0 & 0 & 0 & \ldots & a_{-l+1} & 0 \end{bmatrix} \tag{12.7}$$

where by Eq. (12.4)

$$a_{l-n} = \sqrt{(2l - n)(n + 1)}$$

Hence the operator L_x is said to be nondiagonal in the spherical coordinate representation with the z axis parallel to the polar axis. This is because (see below) L_z is diagonal in this representation and L_x and L_y do not commute with L_z. In general, only operators that commute can both be diagonal in the same representation; conversely if a representation exists in which two operators are diagonal these two operators commute. Since the energy of a complete system is constant and physically observable, a representation can always be found in which it is diagonal. In Chapter 10 it was pointed out that the representation in which the energy operator was diagonal was called the true or correct representation and the wave functions in this representation were called the true or correct functions of the system.

We can easily demonstrate that two noncommuting operators P and Q cannot both be diagonal in the same representation. We ask, what is $Q\psi$ if $P\psi = p\psi$ and $[P, Q] \equiv PQ - QP = R$, the capital letters denoting operators and p denoting the eigenvalue? To determine the answer to this question we operate on $Q\psi$ with P:

$$P(Q\psi) = PQ\psi = (QP + R)\psi = pQ\psi + R\psi$$

Thus P operating on $Q\psi$ does not give the same eigenvalue and unchanged eigenfunction as are obtained with P operating on ψ. Q changes the wave function (and hence must be nondiagonal) so that $P(Q\psi)$ gives a different result, even within a factor of the eigenvalue of Q, from that given by $P\psi$. The converse of this statement, incidentally, is not true. It does not follow, if P commutes with Q, and if P is diagonal, that Q is also diagonal. This is shown by a simple example. The operator L^2 is diagonal in the present representation (with diagonal elements $l(l+1)$), and it commutes with L_x, which is not diagonal.

12.2   SPIN

The energy of a magnet in a magnetic field depends on its orientation in the field and is given by $\mathbf{M} \cdot \mathbf{B}$ where $\mathbf{M}$ is the magnetic moment of the magnet. An electron of mass m_e having an angular momentum L_z about an atomic nucleus would give the atom a magnetic moment μ of $L_z e/2m_e c$ ergs per gauss because of its charge. Since the projection of the electron angular momentum on any axis is quantized, the product $\mu \cdot \mathbf{B}$ will also be quantized and the energy of the atom in a magnetic field directed along the z axis will be given by

$$E = m(e\hbar/2m_e c)B_z \qquad (12.8)$$

where m is the magnetic quantum number of the electron. If the atom is in an inhomogeneous magnetic field, one "pole of the magnet" may be in a

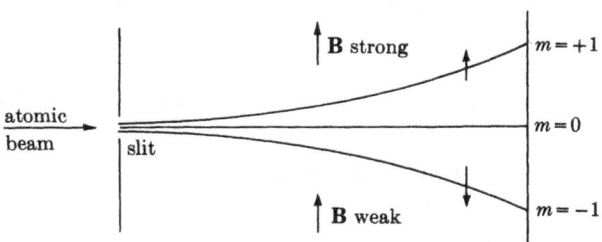

Figure 12.1 *Illustration of the deflection of a beam of atoms having one unit of angular momentum in an inhomogeneous magnetic field.*

stronger magnetic field than the other and thus the total magnet (the atom) will experience a force (in addition to a torque) given by

$$F_z = \frac{dE}{dz} = \frac{\partial E \partial B_z}{\partial B_z \partial z} = \frac{me\hbar}{2m_e c}\frac{\partial B_z}{\partial z}$$

Stern and Gerlach utilized this effect in a historic experiment similar to the experiment illustrated in Figure 12.1 for a beam of atoms having one unit of total angular momentum. Three spots would be found on the screen at the right rather than a single broad smudge, showing that the projection of the angular momentum on the magnetic field is quantized.

The quantized projection of the angular momentum on a given axis, illustrated by this experiment, can have $2l + 1$ discrete values where l is the total orbital momentum of the atoms corresponding to the values $-l, -l + 1, ..., l - 1, l$ which the magnetic quantum number may assume. Hence, integrality of l requires an odd number of spots in a Stern–Gerlach experiment. For hydrogen, for example (or for the silver atoms Stern and Gerlach used in their historic experiment) only one spot would be expected since the electron in the ground state has no orbital angular momentum. *However, two spots are observed!* For $l = 1$, four spots are observed for these atoms. An even multiplicity for m necessarily implies half-integral total angular momentum, since the multiplicity of m is $2l + 1 = (2(\frac{1}{2}) + 1)$ if $l = \frac{1}{2}$; in this regard G. Uhlenbeck and S. Goudsmit suggested that the electrons themselves had an inherent angular momentum or *spin* of one half unit (whose eigenvalue, nevertheless, is $\hbar\sqrt{\frac{1}{2}(1 + \frac{1}{2})} = \sqrt{3}\hbar/2$). This inherent angular momentum is quantized like any other angular momentum. Its projection on any axis, including the axis of the orbital angular momentum, is $\pm\frac{1}{2}\hbar$.

A large homogeneous magnetic field will break up the $2l + 1$ degeneracy in magnetic quantum number m occurring for atomic energy levels. The splitting of atomic levels obtained by this means in atomic bright-line spectra is called the Zeeman effect. Equation (12.8) gives the dependence of the electron energy on the orientation of the total electron angular momentum in the externally applied magnetic field. Similar to the Stern-Gerlach experiment, the multiplicity of the levels in the bright-line spectra is observed to be $(2j + 1)$ where j is odd-half-integral for atoms having an odd number of electrons, giving an even multiplicity for these atoms. The strength of the magnetic dipole per unit angular momentum due to the electron spin is experimentally observed to be twice as great as given in Eq. (12.8) for a charge having a z-projected orbital momentum of $m\hbar$ units; inclusion of this correction, along with the assumption of spin, enables one to calculate correctly all the degeneracy and magnetic interaction effects which electrons have been observed to undergo.

Another important magnetic interaction of atomic electrons is the spin–orbit interaction discussed in Chapter 9. Since if $l \neq 0$ an atomic electron is necessarily in motion about the nucleus, the static electric field of the nucleus appears as a magnetic field as well, when relativistically transformed to the velocity frame of the electron. If V is the nuclear electrostatic potential in which the electron moves, then

$$\mathbf{B} = \frac{1}{c} \mathbf{v} \times \mathbf{E} = \frac{1}{c} \mathbf{v} \times \left(-\frac{\mathbf{r}}{r} \frac{\partial V}{\partial r} \right)$$

$$= \frac{1}{cr} \frac{\partial V}{\partial r}(\mathbf{r} \times \mathbf{v}) = \frac{1}{cr} \frac{\partial V}{\partial r}\left(\frac{1}{m_e} \mathbf{L} \right) \tag{12.9}$$

where $\mathbf{L}$ is used, as always, to denote the orbital angular momentum vector operator. The magnetic energy of the spinning electron in the magnetic field created by the nuclear electrostatic field and the electron's orbital motion is

$$E = \frac{e}{4m_e^2 c^2} \frac{1}{r} \frac{\partial V}{\partial r} \mathbf{L} \cdot \mathbf{S} \tag{12.10}$$

where $\mathbf{S}$ denotes the angular momentum of the electron spin. (Note that, owing to relativistic corrections calculated by L. H. Thomas, E in Eq. (12.10) is one half the result of combining Eqs. (12.8) and (12.9).)

The expression (12.32) was used in Eq. (9.34) in estimating the spin–orbit interaction for atomic electrons. Thus, the spin–orbit interaction also splits up the $2l + 1$ degeneracy in m (i.e., the $2l + 1$ different values of m resulting in $2l + 1$ different states with the same energy), giving rise to fine structure in atomic spectra, that is, groups of closely spaced lines in highly resolved spectra where, in the absence of the spin–orbit effect, only one line would be observed.

The Dirac relativistic quantum theory for electrons, which is beyond the scope of this book, includes the appropriate electron spin and magnetic moment as a necessary consequence of the fundamental theory, rather than as a semi-empirical assumption to explain experiment as the foregoing development suggests.

The electron spin is an additional variable which must be included in the total electron wave function. One way the spin may be represented is by a label telling whether the spin projection on the z axis is up or down (e.g., ψ_+ or ψ_-). A customary notation is to label a spin-up state α and a spin-down state β. When one considers two-electron wave functions, four spin states are possible: $\alpha(1)\alpha(2)$, $\alpha(1)\beta(2)$, $\beta(1)\alpha(2)$, and $\beta(1)\beta(2)$ where 1 and 2 label the two electrons. The total wave function for a spin state in which both spins are projected up would be

$$\psi(\mathbf{r}_1, \mathbf{r}_2)S(1, 2) = \psi(\mathbf{r}_1, \mathbf{r}_2)\alpha(1)\alpha(2) \qquad (12.11)$$

where the two-electron spin eigenfunction has been denoted as $S(1, 2)$. In general, $\alpha^*(1)\alpha(1) = 1$ and $\alpha^*(1)\beta(1) = 0$, i.e., the spin eigenfunctions are in general orthonormal states.

12.3 THE PAULI THEORY OF SPIN

We have not yet presented operators corresponding to the orbital angular momentum operators L_x, L_y, and L_z which will give us the spin of a complete electron state. It is true that we have used an operator of this kind in calculating the spin–orbit energy of an electron (Eq. 12.10), but in our previous discussion we managed to avoid specifying just what it was.

The spatial angular momentum is a property of the state of a system; if the state is degenerate in the projection of its angular momentum on any axis, so that the values of l do not specify all the levels of equal energy, then the state may be represented by a unit-rank tensor (a column tensor or vector) with $2l + 1$ components made up of the $2l + 1$ degenerate wave functions (Eq. 12.5). In the case of half-integral spin only two independent spin wave functions are possible, one representing a projected spin up, the other spin down. The spin eigenfunction of a single electron is either α or β. If we treat the spin eigenfunction as a two-component vector, we have

$$\alpha = \begin{vmatrix} 1 \\ 0 \end{vmatrix} \equiv \text{spin up}, \qquad \beta = \begin{vmatrix} 0 \\ 1 \end{vmatrix} \equiv \text{spin down} \qquad (12.12)$$

The operator representing the spin variable would then be a 2×2 matrix

$$\sigma_i = \begin{bmatrix} a_{11} & a_{12} \\ a_{21} & a_{22} \end{bmatrix} \qquad (12.13)$$

where $i = x$, y, or z and the a's are yet to be evaluated constants. We normalize σ_i to give an eigenvalue of ± 1 so that

$$\sigma_x{}^2 = \sigma_y{}^2 = \sigma_z{}^2 = \begin{vmatrix} 1 & 0 \\ 0 & 1 \end{vmatrix} \equiv 1 \qquad (12.14)$$

where 1 is the unit matrix. $\boldsymbol{\sigma}$ is a vector of constant length and has unit projection on any axis. For the component of $\boldsymbol{\sigma}$ along an arbitrary axis with direction cosines $\gamma_x, \gamma_y, \gamma_z$, we have

$$(\gamma_x \sigma_x + \gamma_y \sigma_y + \gamma_z \sigma_z)^2 = 1$$

which means that

$$\sigma_x \sigma_y + \sigma_y \sigma_x = \sigma_x \sigma_z + \sigma_z \sigma_x = \sigma_y \sigma_z + \sigma_z \sigma_y = 0 \qquad (12.15)$$

that is, the σ_i's *anticommute*. We might note here that the operator $\boldsymbol{\sigma}^2$ is *not* a unit operator even though it has unit projection on any axis (see Problem 12.1). Arbitrarily we decide on a representation in which σ_z is diagonal and is different from the unit matrix. Then

$$\sigma_z = \begin{vmatrix} a & 0 \\ 0 & b \end{vmatrix}$$

We have $\sigma_z \alpha = \alpha$, $\sigma_z \beta = -\beta$. Therefore $a = 1$, $b = 1$, and

$$\sigma_z = \begin{vmatrix} 1 & 0 \\ 0 & -1 \end{vmatrix} \qquad (12.16)$$

which satisfies Eq. (12.14).

In Chapter 5 we showed that real physical observables can be represented only by Hermitian operators. The corresponding property for matrices is $a_{nk}{}^* = a_{kn}$, as is readily verified by writing

$$a_{nk}{}^* = [\int \varphi_n{}^* A \varphi_k d\tau]^* = \int \varphi_n (A\varphi_k)^* d\tau = \int \varphi_k{}^* A \varphi_n d\tau = a_{kn}$$

That is, the complex conjugate of the transposed matrix must be equal to the original matrix, $a_{nk}{}^* = a_{kn}$. Hence, the diagonal elements must be real and

$$\sigma_x = \begin{bmatrix} a_{11} & a_{12} \\ a_{12}{}^* & a_{22} \end{bmatrix}$$

From Eq. (12.15), $\sigma_x \sigma_z + \sigma_z \sigma_x = 0$, so that multiplying matrices gives us

$$\begin{bmatrix} a_{11} & -a_{12} \\ a_{12}{}^* & -a_{22} \end{bmatrix} + \begin{bmatrix} a_{11} & a_{12} \\ -a_{12}{}^* & -a_{22} \end{bmatrix} = \begin{bmatrix} 2a_{11} & 0 \\ 0 & -2a_{22} \end{bmatrix} = 0$$

Therefore $a_{11} = a_{22} = 0$. From Eq. (12.14),

$$\sigma_x{}^2 = \begin{bmatrix} |a_{12}|^2 & 0 \\ 0 & |a_{12}|^2 \end{bmatrix} = \begin{bmatrix} 1 & 0 \\ 0 & 1 \end{bmatrix}$$

Thus $|a_{12}|^2 = 1$, which means that the magnitude (or modulus) of a_{12} is unity, but since a_{12} is a complex number its phase, μ, is as yet undetermined; therefore $a_{12} = e^{i\mu}$. If we choose $\mu = 0$, we obtain for this component of the spin operator

$$\sigma_x = \begin{bmatrix} 0 & 1 \\ 1 & 0 \end{bmatrix} \tag{12.17}$$

The same relations hold for σ_y. In order for σ_y to anticommute with σ_x, we must have $\mu = \pm \pi/2$. We choose $-\pi/2$ and obtain

$$\sigma_y = \begin{bmatrix} 0 & -i \\ i & 0 \end{bmatrix} \tag{12.18}$$

The spin of the electron is given by the operator $\mathbf{S}$,

$$\mathbf{S} = (\hbar/2)\boldsymbol{\sigma} \tag{12.19}$$

so that

$$S_z \alpha = S_z \begin{vmatrix} 1 \\ 0 \end{vmatrix} = \frac{\hbar}{2} \begin{vmatrix} 1 \\ 0 \end{vmatrix} = \frac{\hbar}{2} \alpha \tag{12.20}$$

since by straight matrix multiplication $\sigma_z \begin{vmatrix} 1 \\ 0 \end{vmatrix} = \begin{vmatrix} 1 \\ 0 \end{vmatrix}$. That is, the eigenvalue of the spin operator S_z is $\hbar/2$ when S_z operates on α. Similarly

$$S_z \beta = -\frac{\hbar}{2} \beta \tag{12.21}$$

and

$$S_x \alpha = \frac{\hbar}{2} \beta \quad \text{and} \quad S_x \beta = \frac{\hbar}{2} \alpha \tag{12.22}$$

Note, however, that

$$S_x(S_x \alpha) = S_x \left(\frac{\hbar}{2} \beta\right) = \frac{\hbar^2}{4} \alpha$$

as it should since the operator $\sigma_x{}^2$ equals the unit matrix. $\mathbf{S}$ is an operator representing the spin or intrinsic angular momentum of the electron, and it operates on the spin eigenfunctions (not on space!). The operator S^2 has the eigenvalue $\frac{3}{4}\hbar^2$. If we include the $\mathbf{L} \cdot \mathbf{S}$ term in the Hamiltonian (the H operator) for atomic electrons, then if we use central field potential wave functions, H is not diagonal, for $\mathbf{L} \cdot \mathbf{S}$ does not commute with L_z or S_z, which are diagonal in this representation. Operation with the $\mathbf{L} \cdot \mathbf{S}$ term will mix states of different L_z, S_z (since the $\mathbf{L} \cdot \mathbf{S}$ operator has nondiagonal matrix elements) having the same $m = m_l + m_s$, as we have discussed in the section on fine structure in Chapter 9 (Eqs. 9.34–9.36). m_l and m_s are the z-axis projections of the orbital and spin angular momentum vectors respectively. The operator $\mathbf{L} \cdot \mathbf{S}$ was found to have the eigenvalues $(\hbar^2/2)l$ for $j = l + s = l + \frac{1}{2}$ and $(-\hbar^2/2)(l + 1)$ for $j = l - \frac{1}{2}$ in this representation. Table 12.1 and Eq. (12.49) in Section 12.6 give the transformation matrix from the $|m_l, m_s\rangle$ to the $|j, m\rangle$ wave functions in the diagonal $(|j, m\rangle)$ representation where in this instance the subscript 1 in the table is l and 2 is s.

Problem 12.1   Prove that

$$\boldsymbol{\sigma}^2 = 3, \quad \sigma_x\sigma_y = i\sigma_z, \quad \boldsymbol{\sigma} \times \boldsymbol{\sigma} = 2i\boldsymbol{\sigma}$$

Problem 12.2   Show that the operator

$$e^{\sigma_y} = \cosh 1 + \sigma_y \sinh 1$$

Problem 12.3   Show that $\mathbf{L} \cdot \mathbf{S}$ does not commute with L_z or S_z but that it does commute with J_z, where $J_z = L_z + S_z$.

12.4 SYMMETRY

For a system containing two identical particles, one cannot distinguish between two eigenfunctions which differ only in that the particles are interchanged. They are both eigenfunctions of the same eigenvalue. For example, if x denotes spin and space coordinates, then in

$$H\psi(x_1, x_2) = E\psi(x_1, x_2)$$

$$H\psi(x_2, x_1) = E\psi(x_2, x_1)$$

$\psi(x_1, x_2)$ and $\psi(x_2, x_1)$ are degenerate eigenfunctions of the H operator. The exchange operation consists in exchanging the two particles, and when ψ

is an eigenfunction of the exchange operator

$$P\psi(x_1, x_2) = k\psi(x_2, x_1)$$

where k is the eigenvalue of the exchange operator. When the exchange operation is taken twice, one must wind up with the original eigenfunction:

$$P^2\psi(x_1, x_2) = k^2\psi(x_1, x_2) = \psi(x_1, x_2)$$

Therefore $k = \pm 1$, whence

$$P\psi(x_1, x_2) = + \psi(x_2, x_1)$$

or

$$P\psi(x_1, x_2) = - \psi(x_2, x_1)$$

the first equation defining a symmetric eigenfunction and the second an antisymmetric eigenfunction. If the state is degenerate and two or more eigenfunctions are possible for a given eigenvalue, it is possible to construct combinations of these eigenfunctions which are either entirely symmetric or entirely antisymmetric eigenfunctions, as will be shown.

The symmetry properties of the spin eigenfunction are of crucial importance. If particles 1 and 2 are interchanged in either the $\alpha(1)\alpha(2)$ or the $\beta(1)\beta(2)$ state, the eigenfunction is unchanged and therefore these states are said to be *symmetric*. The other states $\alpha(1)\beta(2)$ and $\beta(1)\alpha(2)$ are neither entirely symmetric nor entirely antisymmetric, however, and it is convenient to treat them as a superposition of two other states one of which is entirely symmetric and the other is antisymmetric. We therefore use linear combinations of the $\alpha(1)\beta(2)$ and $\beta(1)\alpha(2)$ states to represent the two remaining basic spin functions of the system:

$$\frac{1}{\sqrt{2}} [\alpha(1)\beta(2) + \beta(1)\alpha(2)] \tag{12.23}$$

and

$$\frac{1}{\sqrt{2}} [\alpha(1)\beta(2) - \beta(1)\alpha(2)] \tag{12.24}$$

Equation (12.23) is entirely symmetric, yielding a third basic symmetric spin state; Eq. (12.24) is entirely antisymmetric.

The projection of the spin of the $\alpha(1)\alpha(2)$ state on the z axis is $+\hbar$, that of $\beta(1)\beta(2)$ is $-\hbar$, and that of the other two states (one of which is symmetric, one of which is antisymmetric) is zero. This corresponds to the three values that the projection of unit intrinsic angular momentum can have in a triplet state in which all three eigenfunctions are symmetric, and to the single projection value that the zero total spin of an antisymmetric singlet state can have. That is, the spins of the two electrons are parallel in the three symmetric states (collectively referred to as the triplet state) and antiparallel in the singlet state.

As we have seen in Chapter 11, Eqs. (11.40–11.42) and Eq. (11.45), the two-electron space wave functions may also be symmetric or antisymmetric. (The reader may remember, that in the discussion of quantum mechanical resonance the correct wave functions, Eqs. (11.40–11.42, 11.45), were symmetric and antisymmetric space wave functions, i.e., if the electrons are exchanged, their space wave functions do or do not change sign.) The total two-electron eigenfunction will be symmetric if the space and spin states are both symmetric or both antisymmetric; it will be antisymmetric if the space and spin states have differing symmetry. *It is a fundamental postulate of physics, called the Pauli exclusion principle, that two identical particles having odd half-integral spin* ($\pm \frac{1}{2}$, $\pm \frac{3}{2}$, *etc.*) *must occupy antisymmetric states.* No exception to this principle has ever been found. Two identical half-integral spin particles occupying the same space state must obviously have a total space state which is symmetric, and so their spin state must be the antisymmetric function (Eq. 12.24). Identical particles having integral spin (0, ± 1, ± 2, etc.) occupy symmetrical states. It is not possible, however, for three particles to be in an antisymmetric spin state. Hence for an antisymmetric three-electron wave function, as in the case of the lithium atom, the third electron must exist in a state which is antisymmetric in space but symmetric in spin to the other two. We might note further that more complex particle systems obey the same symmetry rules. For example, the hydrogen atom is symmetric with respect to other hydrogen atoms since the electron and proton spins add up to a whole integer. Alpha particles and deuterium ions are also symmetric in this respect, but the deuterium atom, the tritium ion, and the N^{14} atom are antisymmetric.

For many-electron atoms, a Slater determinant wave function may be formed. This is a superposition of product wave functions of the individual electrons which makes use of the fact that the interchange of any two columns of a determinant (corresponding to the operation of exchanging two electrons) changes the sign of the determinant. With χ denoting the spin of the electron and φ the electron total space and spin state, the determinant is written

$$\psi = \mathrm{Det} \begin{vmatrix} \varphi_1(\mathbf{r}_1, \chi_1) & \varphi_1(\mathbf{r}_2, \chi_2) & \cdots & \varphi_1(\mathbf{r}_n, \chi_n) \\ \varphi_2(\mathbf{r}_1, \chi_1) & \varphi_2(\mathbf{r}_2, \chi_2) & \cdots & \varphi_2(\mathbf{r}_n, \chi_n) \\ \cdot & \cdot & & \cdot \\ \cdot & \cdot & & \cdot \\ \cdot & \cdot & & \cdot \\ \varphi_n(\mathbf{r}_1, \chi_1) & \varphi_n(\mathbf{r}_2, \chi_2) & \cdots & \varphi_n(\mathbf{r}_n, \chi_n) \end{vmatrix} \qquad (12.25)$$

It was noted in Chapter 11 that in atoms antisymmetric electron space wave functions yield lower energies than the symmetric electron space wave functions. This may be interpreted physically as resulting from the repulsive force between the electrons: The antisymmetric space wave functions overlap much less than the symmetric space wave functions, which makes the Coulomb repulsion between the two electrons less in the antisymmetric than in the symmetric space state. Thus, the Pauli exclusion principle leads, in general, to an indirect dependence of the electron energies on their spin state. Since electrons have half-integral spin, the total electronic wave function is required to be antisymmetric, and space-antisymmetric electron states are lower in energy than space-symmetric ones; therefore among degenerate space states the corresponding symmetric spin states will be lower in energy than the antisymmetric spin states. Another way of saying this is Hund's rule, *states of highest spin will have the lowest energy,* for the antisymmetric two-electron spin state is composed of electrons whose spins couple to zero, and the symmetric two-electron spin state is composed of electrons whose spins couple to one. In general, for n electrons, the state with z projection of spin $n\hbar/2$ has a total spin of $n\hbar/2$. This maximum value of z projection of spin, $n\hbar/2$, can occur only if all n electrons are lined up in one direction along the z axis with spin up, i.e., all electrons are in the same spin state α. This state is necessarily symmetric in spin, and multiplies a necessarily totally antisymmetric space wave function. Similarly, the state with z projection of spin $(n - 1)\hbar/2$ has $n - 1$ electrons in an α state and one electron in a β state. This state is composed *partly* of symmetric spin states $(\alpha \, \alpha \, \alpha \ldots (\alpha\beta + \beta\alpha))$ which are eigenfunctions of total spin $n\hbar/2$ and multiply a necessarily totally antisymmetric space wave function; however, this state also includes spin states which are eigenfunctions of total spin $(n - 1)\hbar/2$, $\alpha \, \alpha \, \alpha \ldots (\alpha\beta - \beta\alpha)$. Hence these constituent spin states are spin-antisymmetric with respect to the exchange of two electrons and spin-symmetric with respect to any other electron pair. Since the space state must be symmetric with respect to the electron pair which is antisymmetric in spin, this state with total spin $(n - 1)\hbar/2$ is higher in energy than the state of total spin $n\hbar/2$, in accord with Hund's rule. A state with total spin of $(n - 2)\hbar/2$ is of still higher energy, and so on for the other states. In summary, the space function must be symmetric in $(n/2) - (S/\hbar)$ pairs, where S is the total spin of the wave function. This phenomenon where the spin and energy are indirectly coupled is called Russell–Saunders coupling, which we discussed less generally in Section 11.0. In nuclear interactions the reverse effect occurs, for the nuclear force, in contrast to electron mutual repulsion, is attractive; indeed, the nuclear force, in contrast to the Coulomb force, is itself space symmetry-dependent as well as being attractive and large in a symmetric space state and noticeably smaller in an antisymmetric space state. Therefore, a dominant consideration in determining the arrangement of nucleons in light nuclei is that the total spin of the total nucleon spin eigenfunction be a minimum, for in light nuclei, at least, the total nucleon state with minimum spin is thus the state with least energy, the ground state.

Problem 12.4 Show that $\alpha(1)\alpha(2)$, $\beta(1)\beta(2)$ (Eqs. 12.23, 12 24) are orthonormal eigenfunctions if $\alpha(1)$, $\alpha(2)$, $\beta(1)$, and $\beta(2)$ are normalized and orthogonal to each other.

12.5 PARITY

Parity refers to whether or not the wave function changes sign when the coordinate system is reversed. ψ itself is not a physical observable although $\psi^*\psi$ is, as we know. The change of sign of ψ when the coordinate system is reversed is possible and, as we shall see, is a very useful property of the eigenfunction. It is also convenient to note that the parity of a state may be inferred from the mathematical properties of the function representing the state. We shall also see that operators representing physical observables can be constructed which will tell us whether the parity of a system's eigenfunction *changes* even though the parity itself is not directly observed. In most physical processes this does not happen (i.e., parity is usually *conserved*), but when it does happen that the parity of the system's wave function changes during a physical process, parity is said to be *not conserved*.

The parity operation is equivalent to the exchange operation for systems composed of two identical particles; its application, however, includes also systems of nonidentical particles. The parity operator changes the signs of directions of all coordinates. That is, all vectors $\mathbf{r}_1$, $\mathbf{r}_2 \ldots$ become $-\mathbf{r}_1$, $-\mathbf{r}_2 \ldots$ by a change in the directions of the coordinate system, $x \to -x$, $y \to -y$, $z \to -z$. (This amounts to going from a right-handed to a left-handed coordinate system, which reflection in any plane would also accomplish.) The behavior of the wave function under the parity operation is an important property of the wave function. Since only $|\psi|^2$ is measured experimentally, not ψ itself, the phase of ψ may be affected by the parity operation and the parity operator P may have an eigenvalue different from $+1$. As for the exchange operation, to which the parity operation is equivalent for systems composed of two identical particles, we have

$$P\psi(\mathbf{r}) = k\psi(-\mathbf{r}) \tag{12.26}$$

When the parity operation is taken twice, the original eigenfunction must result:

$$P^2\psi(\mathbf{r}) = k^2\psi(\mathbf{r}) = \psi(\mathbf{r}) \tag{12.27}$$

Therefore $k = \pm 1$; if k is $+1$, the parity of the state is said to be even; if -1, it is said to be odd. Note that the spin coordinate is unaffected by the parity operation. For the *orbital* angular momentum we have

$$L_x = yP_z - zP_y$$
$$\to (-y)(-P_z) - (-z)(-P_y) \tag{12.28}$$
$$= L_x, \text{ not } -L_x$$

and the component of $\mathbf{S}$ must behave in the same way.

If any operator does not depend on time explicitly, its time dependence is determined from its commutation with the Hamiltonian (Eq. 5.17):

$$[P, H] = PH - HP \qquad (12.29)$$

Note that this result shows that parity is conserved (unchanged) for all space-symmetric Hamiltonians, for H being space-symmetric does not change when operated on by P. Since $(PH) = H$, $PH = HP$, and therefore $PH - HP = 0$. ((PH) means P operates only on H while PH means P operates on *everything* to the right of P.) The consequence, $dP/dT = 0$, means that if the initial state of a system is even (or odd), it will remain so in the course of time. The parity of the state will never change. Formerly it was thought that all systems have Hamiltonians that commute with the parity operator, but as we shall presently see this is now known to be false.

The parity of a theoretical wave function is readily determined by reversing the coordinate system and calculating the new wave function. Through such calculations it becomes clear that in the case of a spherical harmonic wave function, the parity is equal to $(-1)^l$, where l is the total orbital momentum quantum number. Therefore, s, d, and g states ($l = 0$, 2, 4) have even parity ($k = 1$), while p, f, and h states ($l = 1, 3, 5$) have odd parity ($k = 1$).

The parity of the total state of an entire system is not directly determinable experimentally; it must be inferred from a measurement of the total orbital momentum of the state. Let us consider in detail what we can and cannot learn about the parity of a state through physical measurement. An even parity operator representing an observable, O_+, would yield the result

$$\langle \psi^* | O_+ | \psi \rangle = \langle \psi_+^* | O_+ | \psi_+ \rangle + \langle \psi_-^* | O_+ | \psi_- \rangle + \langle \psi_+^* | O_+ | \psi_- \rangle + \langle \psi_-^* | O_+ | \psi_+ \rangle$$
$$(12.30)$$

for a ψ impure in parity, being composed of ψ_+ an even-parity and ψ_- an odd-parity wave function. That is,

$$P\psi_+(\mathbf{r}) = \psi_+(-\mathbf{r}) \quad \text{and} \quad P\psi_-(\mathbf{r}) = -\psi_-(-\mathbf{r}) \qquad (12.31)$$

When the integrals over all space are taken, the cross-product terms $\langle \psi_+^* | O_+ | \psi_- \rangle$ and $\langle \psi_-^* | O_+ | \psi_+ \rangle$ give zero. The other two terms are positive definite. A measurement of the physical quantity represented by O_+ would therefore not yield any information about the parity of ψ. However, the fact that the parity of the state is *not pure* could be determined with an odd-parity operator of a different observable, O_-. An odd-parity operator, which changes sign if the coordinate system is reversed, gives a finite result only for the cross-product terms $\langle \psi_+^* | O_- | \psi_- \rangle$ and $\langle \psi_-^* | O_- | \psi_+ \rangle$. Hence, if the physical observable represented by O_- is measured experimentally and a nonzero result is obtained, we have definite evidence of an impure parity state representing the system on which the measurement is made. On the other hand, if the parity of the state is pure, but changes during some transition to another state and if O_- is somehow measured between the initial state on the right and the final state on the left and gives a nonzero result, we know that the parity of the system changed during the transition.

If the parity of a state were to change during some physical process, say particle emission, a partial change in parity of the system state function during the particle emission process might be observed as an asymmetry in the spatial distribution of the emitted particles. In one famous example, beta decay as initially commented on by C. N. Yang and T. D. Lee, one might observe the correlations or scalar products of electron momentum **p** and the spin (sum of intrinsic nucleon spins plus the nuclear orbital momentum) of the emitting nucleus, **J**, $(\mathbf{p} \cdot \mathbf{J})$; electron momentum and the circular polarization $\boldsymbol{\Sigma}$ (angular momentum) of an emitted gamma ray associated with the beta decay, $(\mathbf{p} \cdot \boldsymbol{\Sigma})$; circular polarization of a gamma ray and its propagation vector, or electron momentum and its intrinsic spin $(\mathbf{p} \cdot \boldsymbol{\sigma})$. Each of these operators includes an angular momentum vector **J**, $\boldsymbol{\Sigma}$, or $\boldsymbol{\sigma}$ which is not a true vector. If the coordinate system is reversed (that is, $x \rightarrow -x$, $y \rightarrow -y$, $z \rightarrow -z$) so that one goes from a so-called right-handed to a left-handed coordinate system, all these scalar products change sign. The axis of this asymmetry would clearly have to have some physical meaning, for all directions in space would be equivalent unless some property of the system had a direction one could choose as a coordinate axis for analyzing the emission event. Such directions or axes with physical meaning are furnished by pseudovectors representing physical variables, e.g., **J**, $\boldsymbol{\sigma}$, or $\boldsymbol{\Sigma}$. Ordinary scalar quantities such as the mass of a sack of potatoes or the temperature of a room never depend on a coordinate system in this way;

therefore, the scalar product of angular momentum and ordinary momentum is called a *pseudoscalar*. A pseudoscalar operator such as $\mathbf{p} \cdot \boldsymbol{\sigma}$ is, therefore, an odd parity operator which represents a physical observable. A positive measurement of correlation between electron momentum $\mathbf{p}$ and electron spin $\boldsymbol{\sigma}$ is proof that an impure parity state exists.

For beta decay, an initial atomic state is represented as transformed to a final state by use of the beta decay operator:

$$\psi_f = H_\beta \psi_i$$

where ψ_f is composed of product wave functions of the residual nucleus, the atomic electrons, the emitted beta ray, and neutrinos, and ψ_i represents the initial atom and H_β the beta decay operator.

The expectation of a pseudoscalar operator, for example $\mathbf{J} \cdot \mathbf{p}$ in beta decay, is given by

$$\langle \mathbf{J} \cdot \mathbf{p} \rangle = \langle \psi_f^* | \mathbf{J} \cdot \mathbf{p} | \psi_f \rangle = \langle \psi_i^* H_\beta^* | \mathbf{J} \cdot \mathbf{p} | H_\beta \psi_i \rangle \tag{12.32}$$

In integration over all space an odd number of odd-parity factors in the integrand will cause the integral to be zero.

Ordinarily, without use of the $\mathbf{J} \cdot \mathbf{p}$ operator, no interference can be observed between the parity-conserving and parity-nonconserving terms in the beta decay operator; the calculated result is proportional to the *square* of these terms and hence their parity cannot be ascertained. The $\mathbf{J} \cdot \mathbf{p}$ operator, however, is an odd parity operator, i.e., it is odd with respect to reversing the coordinates and when integrated gives a finite result only if multiplied by an odd parity function on one side and an even parity function on the other side. Thus $\mathbf{J} \cdot \mathbf{p}$ or any pseudoscalar operator would have an expected value of zero unless the beta decay interaction operator contained both even and odd terms. The presence of both odd and even parity terms was first determined experimentally by C. S. Wu and co-workers at Columbia and the N.B.S., by observing the beta decay electron momentum $\mathbf{p}$ upon emission from nuclei whose nuclear spins $\mathbf{J}$ were aligned all in one direction. The electrons were not emitted isotropically as expected; more electrons were measured in one direction along the axis of nuclear spin than in the opposite direction. The observed correlation of emitted beta rays with nuclear spin was incontrovertible evidence of partial parity nonconservation in the beta decay process. This, and μ meson decay, were the physical interactions or processes first observed not to conserve parity. For nuclear or electromagnetic interactions, such as give rise to alpha particle decay or different nuclear states, operators such as $\mathbf{J} \cdot \mathbf{p}$ are observed to give a zero eigenvalue; the rule appears to be that strong interactions conserve parity and weak interactions such as beta decay do not.

12.6 VECTOR ADDITION COEFFICIENTS

Even though the total spin and orbital momentum of a state may be given, the total angular momentum is not specified thereby. Any two vectors such as the spin and orbital momentum add up vectorially to a vector, in this case the total angular momentum, whose magnitude ranges from the difference of the two vectors to their sum. To illustrate, we will consider a total eigenfunction with total angular momentum $\mathbf{j}$, made up of two component eigenfunctions having angular momenta $\mathbf{j}_1$ and $\mathbf{j}_2$, as illustrated in Figure 12.2. This would represent, for example, the total angular momentum of a state composed of a target nucleus with initial angular momentum $\mathbf{j}_1$ which absorbed a nucleus with angular momentum $\mathbf{j}_2$.

First we note that in spherical coordinates a wave function has two angular degrees of freedom; hence there are two eigenvalues associated with the angular momentum. Two angular momentum vectors such as $\mathbf{l}$ and $\mathbf{s}$ or, more generally, $\mathbf{j}_1$ and $\mathbf{j}_2$ will thus have four eigenvalues characterizing their wave functions, j_1, m_1 and j_2, m_2 (from which the angular eigenfunctions may be deduced). We wish to know how these two vectors add. That is, given a wave function ψ which is a simultaneous eigenfunction of the commuting operators $j_1{}^2$, $j_2{}^2$, j^2, and m, we wish to express it as a linear combination of eigenfunctions of the commuting operators $j_1{}^2$, $j_2{}^2$, m_1, and m_2. This is important because the latter functions are simpler (and in fact already known to us in the case of $\mathbf{l}$ and $\mathbf{s}$ for a single electron), whereas the former are the only ones that can also be energy eigenfunctions (as explained in the section on spin-orbit coupling).

We might express our task as finding out how to go from a representation in which the operators $j_1{}^2$, $j_2{}^2$, $j_{1z} = m_1$ and $j_{2z} = m_2$ are diagonal to a representation in which $j_1{}^2$, $j_2{}^2$, j^2, and $j_z = m$ are diagonal, where $[\mathbf{j} = \mathbf{j}_1 + \mathbf{j}_2$ and $j_z = j_{1z} + j_{2z}]$. Since $j_1{}^2$ and $j_2{}^2$ are the same in both representations, we will denote the wave functions in the two representations by $|m_1, m_2\rangle$ and $|j, m\rangle$ respectively. We will also frequently write $|m_1, m_2\rangle$ as $|j_1, m_1\rangle|j_2, m_2\rangle$. Our task is to find out how to go from the first representation to the second, that is to find out how the vectors $\mathbf{j}_1$ and $\mathbf{j}_2$ add.

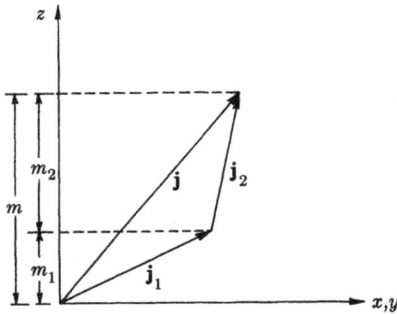

Figure 12.2 *Illustration of vector model of angular momenta, showing how two vectors $\mathbf{j}_1$ and $\mathbf{j}_2$ add to a total vector $\mathbf{j}$.*

For example, a nucleus with known spin may absorb a thermal neutron. The state of the resultant compound nucleus has a total spin j and z projection m and is comprised of various combinations of states of the target nucleus and of the incident neutron state with identical j and m. It is our purpose to determine what these combinations are. The $|j, m\rangle$ may be given as a superposition of the complete set of $|m_1, m_2\rangle$:

$$|j, m\rangle = \sum_{m_1, m_2} a_{jmm_1m_2}|m_1, m_2\rangle$$

Multiplying by an arbitrary state vector $\langle r_1, r_2|$ on the left of both sides we obtain $\langle r_1, r_2|j, m\rangle = a_{jmr_1r_2}$, owing to the orthonormality of the base wave functions $|m_1, m_2\rangle$. Thus, writing m_1 for r_1 and m_2 for r_2 we have

$$|j, m\rangle = \sum_{m_1, m_2} \langle m_1, m_2 | j, m\rangle |m_1, m_2\rangle \qquad (12.33)$$

where $m_1 + m_2 = m$ for each term (since the matrix element vanishes when $m \neq m_1 + m_2$).

The normalized wave function $|j, m\rangle$ is given in the first representation by a linear combination of different $|m_1, m_2\rangle$ whose coefficients are given by the $\langle m_1, m_2 | j, m\rangle$. (Of course, $\langle m_1, m_2 | j, m\rangle = 0$ if $m \neq m_1 + m_2$.)

If the two vectors $\mathbf{j}_1$ and $\mathbf{j}_2$ are parallel and their projection on the z axis is a maximum, then only one compound state is possible:

$$|j, m\rangle = |j, j\rangle = |j_1, j_1\rangle |j_2, j_2\rangle \equiv |j_1, j_2\rangle \qquad (12.34)$$

where $j = j_1 + j_2$ and the third and fourth terms are merely different notations for the same state. If, on the other hand, $m = j_1 + j_2 - 1$, two states are possible,

$$|j_1, j_1\rangle |j_2, j_2 - 1\rangle \quad \text{and} \quad |j_1, j_1 - 1\rangle |j_2, j_2\rangle$$

which may also be written as $|j_1, j_2 - 1\rangle$ and $|j_1 - 1, j_2\rangle$. The complete state $|j_1 + j_2, j_1 + j_2 - 1\rangle$ is a linear combination of these two with appropriate normalized coefficients.

In order to find these coefficients we first note that (cf. the spherical harmonic wave functions introduced in Chapter 8) the operator $j_z(j_x \pm ij_y)$ is

$$j_z(j_x \pm ij_y) = (j_x \pm ij_y)(j_z \pm \hbar) \qquad (12.35)$$

by virtue of the commutation relations for angular momenta. Since

$$j_z|j, m\rangle = m\hbar|j, m\rangle$$

Eq. (12.35) yields

$$j_z(j_x \pm ij_y)|j, m\rangle = (m \pm 1)\hbar(j_x \pm ij_y)|j, m\rangle \qquad (12.36)$$

(Note that eigenvalues, being just numbers, can be written on either side of an operator.) Therefore, $(m \pm 1)\hbar$ is the eigenvalue of j_z (that is, $(m \pm 1)\hbar$ is the projection of the angular momentum on the z axis) when j_z operates on the wave function resulting from the operation of $(j_x \pm ij_y)$ on $|j, m\rangle$. Thus, we conclude that operating on $|j, m\rangle$ with $(j_x \pm ij_y)$ gives us a new eigenfunction $|j, m \pm 1\rangle$ times an as yet undetermined coefficient.

We may determine this coefficient by finding the normalization constant N which goes with the operator $j_x - ij_y$ operating on $|j, m\rangle$ so that both $\langle j, m - 1|j, m - 1\rangle = 1$ and $\langle j, m|j, m\rangle = 1$ (i.e., so that both eigenfunctions are normalized). Since

$$N(j_x - ij_y)|j, m\rangle = |j, m - 1\rangle \tag{12.37}$$

and

$$\langle j, m - 1|j, m - 1\rangle = 1$$

therefore

$$\langle j, m - 1|j, m - 1\rangle = |N|^2\langle j, m|j_x + ij_y\,|j_x - ij_y|j, m\rangle = 1 \tag{12.38}$$

where the complex conjugate of the state $(j_x - ij_y)|j, m\rangle$ has been written on the left of the double line, and N is the normalization constant we seek. Further manipulation utilizing commutativity gives

$$|N|^2\langle j, m|j_x^2 + j_y^2 - i(j_xj_y - j_yj_x)|j, m\rangle$$
$$= |N|^2\langle j, m|j^2 - j_z^2 + \hbar j_z|j, m\rangle$$
$$= |N|^2\hbar^2[j(j + 1) - m(m - 1)] = 1 \tag{12.39}$$

Hence,

$$|N| = \frac{1}{\hbar(\sqrt{j + m})(j - m + 1)} \tag{12.40}$$

(The phase factor in N is arbitrary, since $e^{i\varphi}\,\psi$ represents the same physical state as ψ.)

Therefore, from Eqs. (12.50, 12.53) we conclude that for $m_1 = j_1$,

$$(j_{1x} - ij_{1y})|j_1, j_1\rangle = \hbar\sqrt{(j_1 + j_1)(j_1 - j_1 + 1)}|j_1, j_1 - 1\rangle \tag{12.41}$$
$$= \hbar\sqrt{2j_1}|j_1, j_1 - 1\rangle$$

Similarly, we have

$$(j_x - ij_y)\,|j_1 + j_2, j_1 + j_2\rangle = \hbar\,\sqrt{2(j_1 + j_2)}\,|j_1 + j_2, j_1 + j_2 - 1\rangle \tag{12.42}$$

where the same argument has been applied to the total angular momentum $\mathbf{j} = \mathbf{j}_1 = \mathbf{j}_2$. Now there is only one way in which the eigenfunction $|j_1 + j_2, j_1 + j_2\rangle$ can be formed from eigenfunctions $|j_1, m_1\rangle, |j_2, m_2\rangle$, namely

$$|j_1 + j_2, j_1 + j_2\rangle = |j_1, j_1\rangle\,|j_2, j_2\rangle \tag{12.43}$$

since any other product would necessarily have $m = m_1 + m_2 < j_1 + j_2$. Therefore the left hand side of Eq. (12.42) can be written as

$$(j_{1x} - ij_{1y} + j_{2x} - ij_{2y}) \,|\, j_1, j_1 > \,|\, j_2, j_2 >$$

$$= \hbar \,\sqrt{2j_1}\,|\, j_1, j_1 - 1 > \,|\, j_2, j_2 > \hbar \,\sqrt{2j_2}\,|\, j_1, j_1 > \,|\, j_2, j_2 - 1 >$$

Hence

$$|\, j_1 + j_2, j_1 + j_2 - 1 > \,=\, \sqrt{\frac{j_1}{j_1 + j_2}}\,|\, j_1, j_1 - 1 > \,|\, j_2, j_2 >$$

$$+\, \sqrt{\frac{j_2}{j_1 + j_2}}\,|\, j_1, j_1 > \,|\, j_2 j_2 - 1 > \qquad (12.44)$$

which gives two of the coefficients $< m_1, m_2 \,|\, j, m >$, namely

$$< j_1 - 1, j_2 \,|\, j_1 + j_2, j_1 + j_2 - 1 > \,=\, \sqrt{\frac{j_1}{j_1 + j_2}} \qquad (12.45)$$

$$< j_1, j_2 - 1 \,|\, j_1 + j_2, j_1 + j_2 - 1 > \,=\, \sqrt{\frac{j_2}{j_1 + j_2}} \qquad (12.46)$$

Similarly, from 12.43, we have

$$< m_1, m_2 \,|\, j_1 + j_2, j_1 + j_2 > \,=\, \begin{cases} 0, \text{ unless } j_1 = m_1 \text{ and } j_2 = m_2 \\ 1, \text{ if } j_1 = m_1 \text{ and } j_2 = m_2 \end{cases}$$

All of the other coefficients $< m_1, m_2 \,|\, j, m >$ are obtained by continuing this process, but the calculations get to be very tedious. They are known as the vector-addition coefficients, or the Clebsch-Gordon coefficients, and they have been tabulated extensively.

In general,

$$|\, j, m > \,=\, \sum_{m_1, m_2} \langle m_1, m_2 | j, m \rangle | m_1, m_2 \rangle \qquad (12.33)$$

where the summation must be carried out over all m_1 and m_2 consistent with

$$m = m_1 + m_2$$
$$\mathbf{j} = \mathbf{j}_1 + \mathbf{j}_2 \qquad (12.47)$$

in order to obtain the total wave function $|\, j, m \rangle$. The coefficients on the right-hand side of Eq. (12.33) are frequently written as

$$C_{m_1 m_2 m}^{j_1 j_2 j} = \langle m_1, m_2 | j, m \rangle \tag{12.48}$$

and in many other ways by various authors.

The $C_{m_1 m_2 m}^{j_1 j_2 j}$ are complicated functions of the j_1, j_2, m_1, and m_2, as we have seen, which must be evaluated for each possible combination of j_1 and j_2, m_1, and m_2 occurring in the sum. The C coefficients, called *vector addition coefficients* or Clebsch–Gordon coefficients, occur frequently in atomic and nuclear physics, and extensive tables of possible combinations of m_1, m_2, j_1, and j_2 have been published,* as well as tables for very much more extensive and complicated three-vector addition coefficients for which

$$j = j_1 + j_2 + j_3$$
$$m = m_1 + m_2 + m_3$$

Table 12.1 is a table of vector addition coefficients for the important special case $j_2 = \frac{1}{2}$. With these coefficients one may determine the way the orbital momentum of an electron ($j_1 = l$) may add to its spin ($j_2 = s$) to yield a total angular momentum of j and a z projection of m. If two electrons having a total spin of unity (spin-symmetric state) and a total orbital momentum of l were treated in terms of their total angular momentum and its projection, a 3×3 table having nine coefficients would result.†

What Table 12.1 means in l, s notation is that the spin-space projection wave functions $|m_l, m_s \rangle$ are related to the total angular momentum wave functions $|l \pm \frac{1}{2}, m \rangle$ in the following way:

$$|l + \tfrac{1}{2}, m \rangle = \sqrt{\frac{1}{2} + \frac{m}{2l + 1}} \, |m - \tfrac{1}{2}, \tfrac{1}{2} \rangle + \sqrt{\frac{1}{2} - \frac{m}{2l + 1}} \, |m + \tfrac{1}{2}, -\tfrac{1}{2} \rangle$$

$$|l - \tfrac{1}{2}, m \rangle = -\sqrt{\frac{1}{2} - \frac{m}{2l + 1}} \, |m - \tfrac{1}{2}, \tfrac{1}{2} \rangle + \sqrt{\frac{1}{2} + \frac{m}{2l + 1}} \, |m + \tfrac{1}{2}, -\tfrac{1}{2} \rangle$$

$$\tag{12.49}$$

In matrix notation these wave functions may be written as two-component

---

* See, for example, Condon, E. V., and G. H. Shortley, cited in Bibliography of this chapter.
† Given in Condon and Shortley, *loc cit.*, p. 76.

first-rank tensors or column vectors, and the coefficients in the form of a 2×2 second-rank tensor called a *transformation matrix*, enabling us to go from one representation (the $|m_l, m_s\rangle$) to the other representation ($|j, m\rangle$):

$$
\begin{bmatrix} |l + \tfrac{1}{2}, m\rangle \\[2em] |l - \tfrac{1}{2}, m\rangle \end{bmatrix} = \begin{bmatrix} \sqrt{\dfrac{1}{2} + \dfrac{m}{2l+1}} & \sqrt{\dfrac{1}{2} + \dfrac{m}{2l+1}} \\[2em] -\sqrt{\dfrac{1}{2} - \dfrac{m}{2l+1}} & \sqrt{\dfrac{1}{2} + \dfrac{m}{2l+1}} \end{bmatrix} \begin{bmatrix} |m - \tfrac{1}{2}, \tfrac{1}{2}\rangle \\[2em] |m + \tfrac{1}{2}, -\tfrac{1}{2}\rangle \end{bmatrix}
$$

(12.50)

which gives Eq. (12.49) by the rules of matrix multiplication.

Table 12.1   *Vector addition coefficients for the special case of $j_2 = \tfrac{1}{2}$. Only four different combinations of m_1, m_2, j_1, and j_2 are possible satisfying Eq. (12.47).*

	$m_1 = m - \tfrac{1}{2}$ $m_2 = \tfrac{1}{2}$	$m_1 = m + \tfrac{1}{2}$ $m_2 = -\tfrac{1}{2}$
$j = j_1 + \tfrac{1}{2}$	$\sqrt{\dfrac{1}{2} + \dfrac{m}{2j_1 + 1}}$	$\sqrt{\dfrac{1}{2} - \dfrac{m}{2j_1 + 1}}$
$j = j_1 - \tfrac{1}{2}$	$-\sqrt{\dfrac{1}{2} - \dfrac{m}{2j_1 + 1}}$	$\sqrt{\dfrac{1}{2} + \dfrac{m}{2j_1 + 1}}$

If a spin-1 system were added to the total orbital momentum, three $|j, m\rangle$ wave functions would result, $|l + 1, m\rangle$, $|l, m\rangle$, and $|l - 1, m\rangle$, from the three $|m_l, m_s\rangle$ wave functions $|m - 1, 1\rangle$, $|m, 0\rangle$, and $|m + 1, -1\rangle$. Therefore a 3×3 transformation matrix would be required.

Problem 12.5   Express a total two-electron spin-symmetric state with total orbital momentum l and total angular momentum projected on a given axis of m in terms of the three product wave functions of the two separate electrons, $|m - 1, 1\rangle$, $|m, 0\rangle$, and $|m + 1, -1\rangle$.

BIBLIOGRAPHY

Condon, E. V., and G. H. Shortley, *The Theory of Atomic Spectra*. New York: Cambridge University Press, 1957. Difficult and advanced, this has been a physicist's bible for the material presented in this chapter, since publication of the first edition in 1935.

Slater, J. C., *Quantum Theory of Atomic Structure*, Vol. 1. New York: McGraw-Hill Book Co., 1960. This contains very extensive and especially lucid discussions of angular momenta and the rest of the material presented in this chapter.

Fermi, E., *Notes on Quantum Mechanics*. Chicago: University of Chicago Press, 1961. Fragmentary and unpolished; Fermi never arranged these notes for publication, but they are outstanding for their concise and physically illuminating treatment.

Mayer, M. G., and J. H. D. Jensen, *Elementary Theory of Nuclear Shell Structure*. New York: John Wiley & Sons, Inc., 1955.

ELASTIC SCATTERING THEORY

13

ELASTIC SCATTERING THEORY treats collisions between particles in which no internal excitation of the particles takes place. Inelastic scattering, on the other hand, occurs when the same kind of particle is present in the scattered beam as in the incident beam but the total translational kinetic energy of the system is less, the missing energy going into excitation of internal degrees of freedom of one or both particles. For example, if two atoms collide and part of their translational kinetic energy is lost to electronic excitation or ionization, the scattering is inelastic. Inelastic scattering theory is somewhat more complicated and is outside the scope of this book, so that the internal energy of the system or the complete eigenfunction of the colliding systems' constituent parts will be ignored. A complex, many-component particle such as an atom will be regarded as a single particle.

A cross-section for a given scattering event is defined as the fraction of the number of *incident-particles-per-unit-area* for which the scattering event happens. For example, suppose a group of boys threw 900 baseballs at random through an opening nine square feet in area, producing a beam of 100 balls/square foot. Suppose that out of every 100 balls/foot^2 50 balls hit a target on the other side of the opening. The cross-section would be given by

$$\frac{50 \text{ balls}}{100 \text{ balls/ft}^2} = \tfrac{1}{2} \text{ ft}^2$$

This is the fraction of the incident beam of 100 balls/ft^2 which hit the target on the average. We see that a cross-section is given in units of area and gives the probability that a beam of one ball per square foot will hit the target.

Suppose that 15 balls cracked the target. The cross-section for cracking the target would be

$$\frac{15 \text{ balls}}{100 \text{ balls/ft}^2} = 0.15 \text{ ft}^2$$

It is clear that there are different kinds of cross-sections. We will be concerned solely with the probability of elastically scattering particles into a differential element of solid angle $d\omega$ (the differential elastic scattering cross-section) and the integral of this cross-section over all angles (the total elastic scattering cross-section).

Figure 13.1 illustrates the scattering of a beam of particles by a scattering center. The problem raised is: What is the probability that a particle will be deflected or absorbed and it or a like particle reemitted into a particular differential element of solid angle $d\omega$ if the incident beam has unit intensity (1 particle per cm^2-sec)? The dimension of this probability is, as we have just seen, the dimension of area used to express the incident beam intensity. The scattering cross-section is normally measured in square centimeters, barns, or millibarns where 1 barn $= 10^{-24}$ cm^2. (The name "barn" is said to have been used first in the Manhattan Project during the Second World War, 10^{-24} cm^2 being as huge as a barn compared with most nuclear scattering cross-sections.)

To reiterate, in general the ratio of total number of scattered particles to the number of particles per unit area in the incident beam is called the *total scattering cross-section*. Similarly, the ratio of the number of particles scattered into the differential element of solid angle $d\omega$ to the number of particles per unit area in the incident beam is called the *differential scattering*

cross-section. The number of scattered particles having an energy between
E and $E + dE$ is also of interest, but this is for *inelastic* scattering and outside
the scope of this chapter.

The subject of scattering theory is the relationship of scattering
cross-sections to the potential energy between the interacting particles.
For example, if the wavefunctions of the electrons in an atom are known,
the electrostatic potential energy of an incident ion moving near this atom

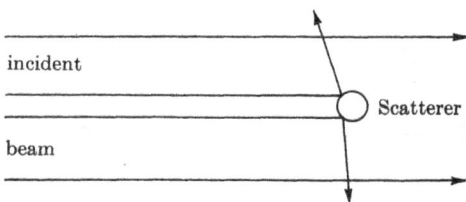

Figure 13.1   *Illustration of the scattering of a beam of particles by a
scattering center.*

can ideally be approximated everywhere as a function of space by ignoring
the changes in atomic wave functions introduced by the ion; scattering theory
then enables us to predict exactly what the differential and total scattering
cross-section of these atoms for the particular incident ions will be. Conversely, detailed measurements of particle scattering from atomic nuclei as
a function of scattering angle and energy of incident and scattered particle
may be interpreted by use of scattering theory to give us a great deal of
information about the potential energy of the scattered particle in the force
field of atomic nuclei. This information is essential to a knowledge of nuclear
forces.

13.1   THE PATH INTEGRAL FORMULATION OF THE
SCATTERING PROBLEM

Before developing the conventional techniques for scattering problems,
in order to obtain a better insight into the scattering process we will first

analyze a simple scattering problem by the path integral method of Chapter 4. That is, we will compute the phase of the exponential in Eq. (4.8) for each path through the scattering center and then integrate over the projected area of the scattering center. As an illustrative example let us take the case of a neutron incident on a completely absorbing nucleus of radius a cm which after absorbing re-emits the neutron at angle θ (i.e., the nucleus

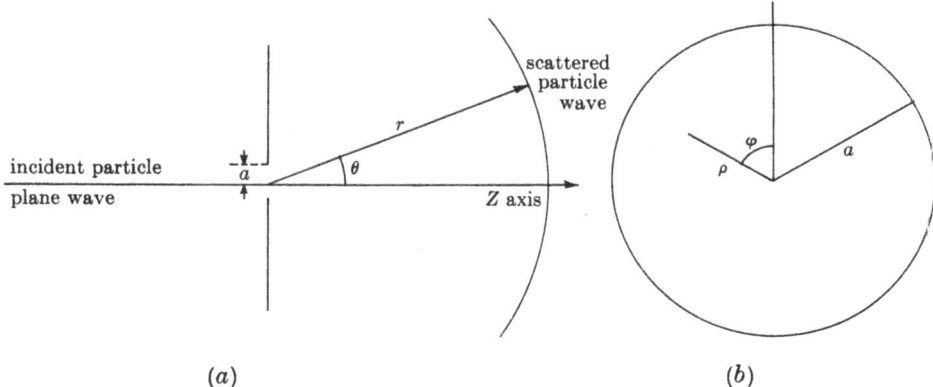

(a)   (b)

Figure 13.2   (a) *Illustration of scattering, showing scattered particle waves. The incident neutron wave is diffracted through the "circular opening" into angle θ. (b) Diagram of circular opening, in plane of the opening, showing a path position ρ, φ at the opening. φ is measured from the plane containing the incident and diffracted wave propagation vectors.*

elastically scatters the neutron). The fraction of the wave stopped by the circular obstruction in a transparent wall may be analyzed in terms of a calculation which treats the nucleus as if it were a two-dimensional opening of a cm radius in an infinite opaque plane perpendicular to the momentum of the incident neutron (see Figure 13.2). The path length x for a path through the point (ρ, φ) in the opening is given by $r - \rho \sin \theta \cos \varphi$, where r is the path length of a ray passing through the center of the opening from the opening to a point r cm away. θ is the scattering angle, and φ is measured from the plane containing the incident and diffracted wave propagation vector. If in Eqs. (4.5, 4.8) we neglect the constant, energy-dependent part, we find

$$\psi(\theta) = Ce^{(i/\hbar)pr} \int_0^a \int_0^{2\pi} \exp\{-(i/\hbar)p\rho \sin\theta \cos\varphi\} \rho \, d\rho \, d\varphi \qquad (13.1)$$

where C is a normalization constant.

Since the integral representation of the Bessel function of order n is

$$J_n(x) = \frac{i^{-n}}{2\pi} \int_0^{2\pi} e^{ix \cos\varphi} e^{in\varphi} d\varphi \qquad (13.2)$$

hence after integrating over φ we are left with

$$\psi(\theta) = Ce^{(i/\hbar)pr} 2\pi \int_0^a J_0\left(\frac{p\rho \sin\theta}{\hbar}\right) \rho \, d\rho$$

and since

$$xJ_0(x) = \frac{d}{dx}[xJ_1(x)]$$

therefore

$$\psi(\theta) = Ce^{(i/\hbar)pr} \frac{2\pi\hbar a}{p \sin\theta} J_1\left(\frac{ap \sin\theta}{\hbar}\right) \qquad (13.3)$$

which yields the amplitude of the scattered wave at angle θ.

The total intensity at angle θ and distance r beyond the nucleus is found by adding the amplitude of this scattered wave to the amplitude of the incident plane wave $e^{(i/\hbar)pz}$ at the point θ,r and taking the square of the absolute value of this sum. Thus, the probability of finding a neutron at a differential solid angle $d\Omega$ at angle θ is given by

$$\sigma(\theta) \, d\Omega = |e^{(i/\hbar)pz} + \psi(\theta)|^2 \, d\Omega \qquad (13.4)$$

In order to evaluate this expression explicitly, $e^{(i/\hbar)pz}$ must be expanded in spherical harmonics, as is done later in this chapter.

When the neutron is not re-emitted with the same energy as the incident neutron (inelastic scattering) or is not re-emitted at all, the ab-sorbed wave still interferes with the transmitted wave, however, so that elastic scattering is still present. The elastic scattering in terms of our $\psi(\theta)$ just calculated assumes a particularly simple form by application of Babinet's theorem for physical optics, i.e.: The diffraction patterns produced by two complementary screens (the complementary screen is opaque where the other screen is transparent and vice versa) are identical except for the central spot, which corresponds to the diffraction angle zero. Hence, the elastic scattering cross-section for a totally absorbing nucleus is the same as for an opaque screen with a transparent opening of the same size.

$$\sigma(\theta)\,d\Omega \;=\; |\psi(\theta)|^2\,d\Omega \;=\; \left[\frac{Cah J_1(ap\sin\theta/\hbar)}{p\sin\theta}\right]^2 d\Omega \qquad (13.5)$$

The total elastic scattering cross-section when neutrons are absorbed by the nucleus is the fraction of the incident beam intensity which is incident on the nucleus, namely πa^2. By evaluating the integral of $\sigma(\theta)$ over all angles and setting the result equal to πa^2, we determine that $C = 2\sqrt{\pi p/\hbar}$ so that

$$\sigma(\theta)\,d\Omega \;=\; 4\pi a^2\,\csc^2\theta\,J_1{}^2\!\left(\frac{ap\sin\theta}{\hbar}\right) d\Omega \qquad (13.6)$$

This is so-called shadow scattering. The absorption of the neutron causes a loss in the incident wave, a "shadow," and the interference of this shadow with the incident plane wave causes a diffraction pattern which is a function of the angle θ. If the neutron is re-emitted with the same energy, the total scattering cross-section will be $2\pi a^2$, as we shall see in Section 13.9.

13.2 GENERAL FORMULATION OF SCATTERING PROBLEMS

The incident beam of particles is represented by a plane wave e^{ikz} propagated along the z axis, where the particle wave number k is the free particle momentum divided by the particle mass m. Figure 13.2(a) illustrates the physical situation. At very large distances beyond the scattering center the incident particle spatial wave function will be composed of two parts, the initial plane wave and a radially diverging outgoing scattered wave.

That is, for r very much larger than the range of the particle potential in the force field of the scattering center,

$$\psi \sim e^{ikz} + (i/r).e^{ikr}f(\theta) \tag{13.7}$$

Actually all we need say is that ψ is made up of the original plane wave plus a small contribution from the scattered wave representing the few particles which are scattered. Since, however, the scattered wave, which will be our prime concern, will have a radially diverging form, it is most convenient to take as the unknown function to be computed $f(\theta)$, the coefficient of the radially diverging wave, rather than the entire expression for the scattered wave. We are interested in the current of particles scattered into the differential element of solid angle $d\omega$, given by

$$v|\psi(\theta)|^2 r^2 d\omega = v|f(\theta)|^2 d\omega \tag{13.8}$$

where the incident particle current is v for a single particle. The differential scattering cross-section of the scattering center, $\sigma(\theta)\, d\omega$, is thus

$$\sigma(\theta)\, d\omega = |f(\theta)|^2 d\omega \tag{13.9}$$

where $f(\theta)$ has the dimensions of a length. For inelastic scattering in which the particle energy and speed changes, Eq. (13.9) would be multiplied on the right-hand side by $v_{\text{fin}}/v_{\text{init}}$.

13.3 THE BORN APPROXIMATION

The Born approximation is the most effective, simplest, and most generally applied method of computing scattering problems. It is essentially an application of time-dependent perturbation theory to the scattering of a plane wave, with the initial and final wave functions both approximately plane waves far from the scattering center. We have already touched on its application to inelastic atomic scattering in Chapter 10. The primary condition for the validity of this approximation is that the perturbing potential energy V between particle and target be much less than the kinetic energy of the scattered particle, E, so that the amplitude of the perturbed wave $\psi' = (e^{ikr}/r)f(\theta)$ be much less than that of the incident plane wave $\psi^0 = e^{ikz}$.

The time-independent Schroedinger equation for the two-particle interaction may be written for the space-dependent wave function alone since the energy of the particles is a constant in the center of mass system; it is

$$\nabla^2 \psi + \frac{2m}{\hbar^2}[E - V(r)]\psi = 0 \qquad (13.10)$$

where m is the reduced mass of the system, $m_1 m_2/(m_1 + m_2)$. We assume $V \ll E$ and $\psi' \ll \psi^0$, where

$$\psi = \psi^0 + \psi' \qquad (13.11)$$

and $\psi^0 = e^{ik_0 z}$ is the plane wave solution of the homogeneous differential equation

$$\nabla^2 \psi^0 + \frac{2m}{\hbar^2}E\psi^0 = 0 \qquad (13.12)$$

Then first-order time-independent perturbation theory, in which $V\psi'$ is neglected, yields

$$\nabla^2 \psi' + \frac{2m}{\hbar^2}E\psi' = \frac{2m}{\hbar^2}V(r)\psi^0 \qquad (13.13)$$

The right-hand side of (13.13) is merely a known function of r; the left-hand side is the homogeneous part of a differential equation in ψ'.

In the theory of electromagnetism one obtains an equation identical in form to (13.13) for the electrostatic potential φ in a medium with charge density ρ (note that both φ and ρ are the Fourier transforms from time to $k = \omega$),

$$\nabla^2 \varphi + k^2 \varphi = -4\pi\rho$$

whose solution by use of Green's theorem gives one the Lienard–Wiechart retarded potential $\varphi(\mathbf{r})$ at the point $\mathbf{r}$ in terms of the source charge density $\rho(\mathbf{r}')$ at a different point $\mathbf{r}'$,

$$\varphi(\mathbf{r}) = -\int \frac{\rho(\mathbf{r}')}{|\mathbf{r} - \mathbf{r}'|} e^{ik|\mathbf{r}-\mathbf{r}'|} d\tau'$$

where the absolute value of the vector difference $\mathbf{r} - \mathbf{r}'$ is its length. We could take over this result complete and apply it to Eq. (13.13), but particularly for students who are unfamiliar with potential theory we will run through an elementary derivation. Let

$$k_0^2 = \frac{2m}{\hbar^2}E = \frac{p_0^2}{\hbar^2}, \qquad U = \frac{2m}{\hbar^2}V \qquad (13.14)$$

then Eq. (13.13) becomes

$$(\nabla^2 + k_0^2)\psi' = U e^{ik_o z} \qquad (13.15)$$

It is convenient to work with the Fourier transform of ψ',

$$A(\mathbf{k}) = \frac{1}{(2\pi)^{3/2}} \int \psi'(\mathbf{r}) e^{-i\mathbf{k}\cdot\mathbf{r}} d\tau \qquad (13.16a)$$

and the inverse transform,

$$\psi'(\mathbf{r}) = \frac{1}{(2\pi)^{3/2}} \int A(\mathbf{k}) e^{i\mathbf{k}\cdot\mathbf{r}} d\mathbf{k} \qquad (13.16b)$$

where $d\mathbf{k}$ is the three-dimensional volume element in momentum space. Further,

$$\nabla^2 \psi'(\mathbf{r}) = \nabla^2 \frac{1}{(2\pi)^{3/2}} \int A(\mathbf{k}) e^{i\mathbf{k}\cdot\mathbf{r}} d\mathbf{k} = -\frac{1}{(2\pi)^{3/2}} \int k^2 A(\mathbf{k}) e^{i\mathbf{k}\cdot\mathbf{r}} d\mathbf{k} \qquad (13.17)$$

Equation (13.15) then becomes

$$\frac{1}{(2\pi)^{3/2}} \int (-k^2 + k_0^2) A(\mathbf{k}) e^{i\mathbf{k}\cdot\mathbf{r}} d\mathbf{k} = U e^{ik_o z} \qquad (13.18)$$

We now multiply both sides of Eq. (13.18) by a particle plane wave function of particular wave number k', $e^{-ik'\cdot\mathbf{r}}$, and write the integral of both sides over coordinate space,

$$\frac{1}{(2\pi)^{3/2}} \int \int A(\mathbf{k}) e^{-i\mathbf{k}'\cdot\mathbf{r}} e^{i\mathbf{k}\cdot\mathbf{r}} (-k^2 + k_0^2) d\mathbf{k} d\tau = \int U(\mathbf{r}) e^{ik_o z} e^{-i\mathbf{k}'\cdot\mathbf{r}} d\tau \qquad (13.19)$$

By making use of the first of the following mathematical identities resulting from the orthogonality of the wave functions

$$\frac{1}{(2\pi)^{3/2}} \int e^{-i\mathbf{k}\cdot\mathbf{r}} e^{i\mathbf{k}'\cdot\mathbf{r}} d\tau = (2\pi)^{3/2} \delta(\mathbf{k} - \mathbf{k}')$$

$$\frac{1}{(2\pi)^{3/2}} \int e^{-i\mathbf{k}\cdot\mathbf{r}} e^{i\mathbf{k}\cdot\mathbf{r}'} d\mathbf{k} = (2\pi)^{3/2} \delta(\mathbf{r} - \mathbf{r}') \qquad (13.20)$$

we find that Eq. (13.19) becomes

$$(2\pi)^{3/2} A(\mathbf{k}')(-k'^2 + k_0^2) = \int U(\mathbf{r}) e^{i(\mathbf{k}_0 - \mathbf{k}') \cdot \mathbf{r}} \, d\tau$$

and Eq. (13.16b) becomes

$$\psi'(\mathbf{r}') = \frac{1}{(2\pi)^3} \int \left\{ \frac{\int U(\mathbf{r}) e^{i(\mathbf{k}_0 - \mathbf{k}') \cdot \mathbf{r}} \, d\tau}{k_0^2 - k'^2} \right\} e^{i\mathbf{k}' \cdot \mathbf{r}} \, d\mathbf{k}' \tag{13.21}$$

Reversing the order of integration in Eq. (13.21) and integrating over $\mathbf{k}'$, we obtain

$$\psi'(\mathbf{r}') = -\frac{1}{4\pi} \int \frac{e^{ik_0|\mathbf{r}' - \mathbf{r}|}}{|\mathbf{r}' - \mathbf{r}|} U(\mathbf{r}) e^{i\mathbf{k}_0 \mathbf{r}} \, d\tau \tag{13.22}$$

which is the expression we sought for ψ' analogous to the Lienard–Wiechart retarded potential in electromagnetic wave theory. In the Born approximation we are interested in the wave function ψ' only in the region far beyond where the potential energy is nonzero, that is, for $r' \gg r$. For this situation a convenient and useful approximation is to let

$$|\mathbf{r}' - \mathbf{r}| = r' - \widehat{\mathbf{r}'} \cdot \mathbf{r} \tag{13.23}$$

where $\widehat{\mathbf{r}'}$ is the unit vector in the direction of $\mathbf{r}'$ (see Figure 13.3). The approximation results from neglecting terms of order r^2/r'^2 and higher in the expansion of the law of cosines:

$$|\mathbf{r}' - \mathbf{r}| = \sqrt{r'^2 + r^2 - 2\mathbf{r}' \cdot \mathbf{r}} \simeq r'\left(1 - \frac{\mathbf{r}' \cdot \mathbf{r}}{r'^2}\right)$$

In the center of mass system of coordinates, the momentum of the elastically scattered particle, $\mathbf{k}$, is equal in magnitude to the momentum of the incident particle and, of course, it has the direction $\widehat{\mathbf{r}'}$. That is,

$$\mathbf{k} = k_0 \widehat{\mathbf{r}'} \tag{13.24}$$

so that from (13.23)

$$k_0 |\mathbf{r}' - \mathbf{r}| \simeq k_0 r' - \mathbf{k} \cdot \mathbf{r} \tag{13.25}$$

Therefore, ψ' of (13.22) may be written approximately as

$$\psi' = -\frac{1}{4\pi}\frac{e^{ik_o r'}}{r'} \int U(\mathbf{r}) e^{i(\mathbf{k_o}-\mathbf{k})\cdot\mathbf{r}} d\tau \tag{13.26}$$

$$= \frac{e^{ik_o r}}{r'} f(\theta)$$

where $|\mathbf{r}' - \mathbf{r}|$ in the denominator has been put equal to r',

$$f(\theta) = -\frac{1}{2\pi}\frac{m}{\hbar^2} \int V(\mathbf{r}) e^{(i/\hbar)\mathbf{q}\cdot\mathbf{r}} d\tau \tag{13.27}$$

and

$$\mathbf{q} = \hbar(\mathbf{k_0} - \mathbf{k}) = (\mathbf{p_0} - \mathbf{p}) \tag{13.28}$$

where $\mathbf{p}$ is the momentum of the particle after scattering. $\mathbf{q}$ is thus the momentum transfer of the scattered particle, that is, the change in momen-

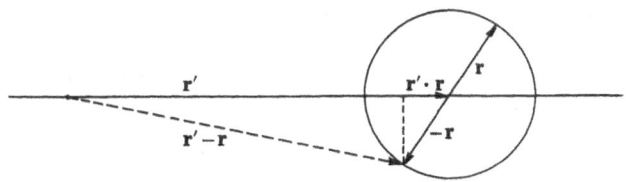

Figure 13.3 *Illustration of the approximation* $|\mathbf{r}' - \mathbf{r}| = r' - \widehat{\mathbf{r}' \cdot \mathbf{r}}$ *which is obtained from the law of cosines for* $r \ll r'$. *$\mathbf{r}'$ is the position of the scattered particle after being scattered in the region* $r \sim 0$.

tum of the particle due to the collision. *Note that $f(\theta)$ is the Fourier transform of the potential energy of the particle interaction.* Thus the result far from the scattering center (13.27) depends only on $\mathbf{q}$ and is completely independent of any other property of the particle. If $V(\mathbf{r}) = V(r)$, a central (angle-independent) potential, $f(\theta)$ is a function of q only, that is, $f(\theta)$ is a function of $|\mathbf{p_0} - \mathbf{p}|$. Since $p = p_0$, the change in momentum is

$$|\mathbf{p_0} - \mathbf{p}| = 2p \sin(\theta/2) \tag{13.29}$$

The particle mass used in the equations throughout this section is the reduced mass $m_1 m_2/(m_1 + m_2)$. If neutrons are scattered from free protons, for example, the reduced mass is one half the nucleon mass. If the protons are bound chemically in an experimental target of polyethylene, however, the effective mass of the target is infinite, i.e., $m_t \gg m_n$ for low energy neutron scattering experiments, and the reduced mass is the neutron mass. This difference of a factor of two in $f(\theta)$ becomes a factor of four in $\sigma(\theta)$ and results in considerably less scattering of neutrons from free protons than from hydrogen compounds.

13.4 SIMPLE APPLICATIONS OF THE BORN APPROXIMATION

1. Square Well If the particle interaction potential is a square well and angle independent, which is a reasonable approximation to low energy nuclear scattering, Eq. (13.27), together with the general result $f(q) = f(\theta)$ from Eqs. (13.28) and (13.29), yields

$$
\begin{aligned}
f(\theta) &= -\frac{m}{\hbar^2} \int_0^a \int_0^\pi V_0\, e^{(i/\hbar)qr \cos\varphi}\, r^2 \sin\varphi\, d\varphi\, dr \\
&= -2\frac{mV_0}{\hbar q} \int_0^a \left(\sin\frac{qr}{\hbar}\right) r\, dr \\
&= \frac{2mV_0\hbar}{q^3}\left[\frac{qa}{\hbar}\cos\frac{qa}{\hbar} - \sin\frac{qa}{\hbar}\right]
\end{aligned}
\tag{13.30}
$$

Although of course low energy nuclear scattering does not meet the validity condition of the Born approximation ($V' \ll E$), Eq. (13.30) is useful in giving a rough idea of the scattering of medium energy neutrons. Since $q = 2p \sin(\theta/2)$ from Eq. (13.29), $f(\theta)$ in Eq. (13.30) may be expressed directly in terms of θ, but it is easier to leave the result in this simple form. Note that $f(\theta)$ is a rapidly decreasing function of q in Eq. (13.30) and is close to a minimum at $\theta_{cm} = \pi$ representing backward scattering in the center of mass system. The differential scattering cross-section $|f(\theta)|^2$ for a square well potential is shown as a function of q in Figure 13.4. For large values of $(qa/\hbar)$, $|f(\theta)|^2 \sim q^{-4} \cos^2(qa/\hbar)$.

2. A Coulomb Potential It is unnecessary to give a quantum mechanical treatment to the Coulomb potential for dissimilar particles of medium or low energy, because, in contrast to the short-range interaction potentials considered thus far, the Coulomb potential, at distances greater than the classical distance of closest approach within the Coulomb and

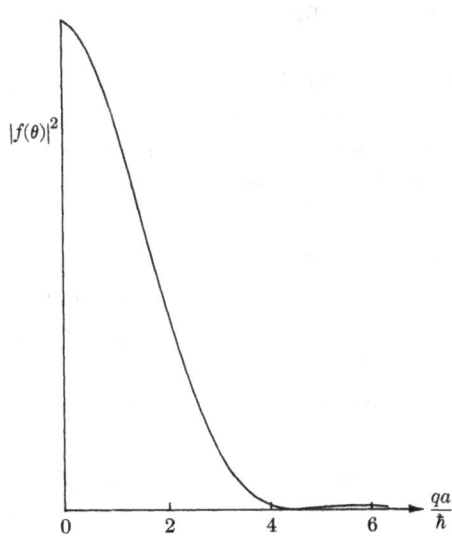

Figure 13.4 *Graph of $|f(\theta)|^2$ for a square well potential as a function of $qa/\hbar$.*

centripetal barriers, changes insignificantly within a particle wavelength. Very little scattering interaction occurs in these classically forbidden regions of space. Therefore the classical calculation by Rutherford is valid. It is always illuminating, however to see how one may obtain classical results by quantum mechanical theory, and a Coulomb potential provides a nice illustration of the scattering to be expected for long-range, $1/r$-dependent potentials. Equation (13.27), the Born approximation for this potential, yields the result

$$f(\theta) = -\frac{mZze^2}{2\pi\hbar^2}\int_0^\infty e^{(i/\hbar)\mathbf{q}\cdot\mathbf{r}}\frac{1}{r}\,d\tau = \frac{-2mZze^2}{q^2} \tag{13.31}$$

whence, by (13.9) and (13.29),

$$\sigma(\theta)\,d\omega = \left(\frac{mZze^2}{2p^2}\right)^2\frac{1}{\sin^4(\theta/2)}\,d\omega \tag{13.32}$$

which is the classical Rutherford formula, in which z is the charge on the incident ion or electron and Z the atomic number of the target atom. Note that in general, $f(\theta)$ is a monotonically decreasing function of the momentum exchange, q, depending inversely on q^2, even for a rather long-range potential. There are interesting quantum mechanical effects occurring in the Coulombic collision of energetic, indistinguishable particles, which will be treated later in this chapter.

3. Constant potential If $V = V_0$, constant over all space, the momentum transform Eq. (13.27) is a delta function of q, $\delta(q)$; thus $f(\theta) = f(q) = 0$ unless $q = 0$, meaning that therefore $q = 0$ in this example, i.e., no momentum transfer takes place: the incident plane wave is transmitted undeflected, as would be expected in a force-free $(\partial V/\partial x = 0)$ situation.

4. Billiard ball collision The interaction between two small hard spheres is essentially a delta function of space. The potential is a square well of nearly infinite height and almost negligible extent. The momentum transform Eq. (13.27) is a constant independent of q and therefore independent of scattering angle θ. Isotropic scattering is approached in a square well at low energies where $V \gg E$ and the particle wavelength $\lambda \gg a$, the range of the potential. However, in this case for which $V \gg E$, the Born approximation becomes rough. The isotropy is an invariable result of $\lambda \gg a$ (see Figure 13.4 for which $qa/\hbar = (2a/\lambda)\sin\theta/2 \ll 1$ for all θ).

Problem 13.2 Why do all instances of inelastic scattering (scattering in which the scattered particle loses energy) include elastic scattering? When elastic scattering is taken into account, what is the maximum total cross-section in terms of the geometrical cross-section of the target?

13.5 QUALITATIVE FEATURES OF THE BORN APPROXIMATION

One can see from these simple problems, and from the discussion of wave packets in Chapter 2 (particularly the example of the Fourier transform of a Gaussian function), that the differential scattering cross-section is, in general, a decreasing function of q. That is, scattering is most likely to occur in directions requiring small change in momentum. Also, since $q = 2p \sin(\theta/2)$ for central potentials, the scattering may be expected to become more pronounced in the forward direction with increasing energy. These are very striking and fundamental features of scattering due to simple, central field interactions. The latter feature is so common and characteristic of all scattering observations that an exception is an exciting and vitally informative discovery in itself. For example, the experimental observation that neutrons scatter backward from protons slightly more than they scatter forward proved that the nuclear force is not a simple central force field. Subsequently this field was found to be best represented by a Serber potential which exchanges half the particles. That is, the nuclear potential is represented most simply as $[\frac{1}{2}(1 + P)]V(r)$ where P is the parity operator and $V(r)$ is a simple central potential. The effect of the parity operator is that an incident neutron emerges as a proton (presumably corresponding to absorption of a positively charged meson from the proton or loss of a negative meson to the proton). In forward exchange scattering, an incident neutron would cause a proton to be scattered at small angles and a neutron at large angles.

13.6 ATOMIC SCATTERING BY THE BORN APPROXIMATION

The potential between an atomic scatterer and an incident ion or electron is that between negatively screened nuclear positive charge and the charge of the incident particle, and may be written by crudely approximating the atom with its many electrons as a simple potential due to only one particle

$$V(r) = ze\left(\frac{Ze}{r} - e\varphi\right) \qquad (13.33)$$

where z and Z are as defined for (13.31) and $-e\varphi$ is the potential resulting from the presence of the target atom electrons. From Poisson's equation in electromagnetic theory, we obtain a differential equation relating the potential to the charge density from which the potential may be computed,

$$\nabla^2(-e\varphi) = -4\pi\rho = -4\pi(-e)n(\mathbf{r})$$

where $n(\mathbf{r})$ is the electron number density (electrons/cc). The nuclear potential is trivial and we have already obtained it, Eq. (13.31). In order to compute the scattering cross-section of an incident charge off an atom we need the atomic shape or the form of the distribution of the atomic electrons. To be specific, we need the atom form factor $F(q)$, which is defined by the integral taken over all space

$$F(q) = \frac{q^2}{4\pi} \int \varphi(r)e^{i\mathbf{q}\cdot\mathbf{r}}\,d\tau \tag{13.34}$$

Let $\chi = -e^{i\mathbf{q}\cdot\mathbf{r}}$; then $\nabla^2\chi = q^2 e^{i\mathbf{q}\cdot\mathbf{r}}$. We now use Green's theorem,

$$\int \nabla^2\chi\,d\tau = \int \nabla\cdot(\nabla\chi)\,d\tau = \int \nabla\chi\cdot d\mathbf{A} = \int \frac{\partial\chi}{\partial n}\,dA$$

where dA is a differential element of surface area and n is in the direction normal to the surface. Since the volume integral is taken over all space, the surface integral over the surface which encloses this volume is that of a sphere of infinite radius. Integrating $F(q)$ twice by parts, we obtain

$$4\pi F(q) = \int \varphi\nabla^2\chi\,d\tau = \int \left(\varphi\frac{\partial\chi}{\partial n} - \chi\frac{\partial\varphi}{\partial n}\right) dA + \int \chi\nabla^2\varphi\,d\tau$$

The surface integral is zero in the limit $r = \infty$, owing to the radial dependences of χ and φ, the atomic potential which must be evaluated on the surface of infinite radius. Therefore

$$F(q) = \int n(r)e^{i\mathbf{q}\cdot\mathbf{r}}\,d\tau \tag{13.35}$$

$F(q)$ may be calculated by use of various atomic electron density approximations such as the simple, inverse power of r-dependent Fermi–Thomas approximation (see Mott and Snedden, Slater, or Pauling and Wilson cited in Bibliography of this chapter). In this way we obtain the result

$$f(\theta) = 2me^2z \frac{Z - F(q)}{q^2}$$

whence

$$\sigma(\theta) = \left(\frac{me^2z}{2p^2}\right)^2 \frac{|Z - F(q)|^2}{\sin^4(\theta/2)} d\omega \qquad (13.36)$$

If q is large, corresponding to large-angle scattering and close approach to the nucleus, Eq. (13.35) gives a nearly zero result due to phase cancellation in the integral over space. Hence if $qa \gg 1$ (where a is the atomic radius), simple Rutherford scattering results. The atomic radius for all atoms is $a \sim a_0 Z^{-1/3}$ where a_0 is the Bohr radius of the hydrogen atom, 0.53 Å. For an alpha particle of 5 Mev, incident on an atom of radius a, for example, $qa \sim 10^4\theta$.

As $q \to 0$,

$$Z - F(q) \to q^2$$

even though Z alone is a constant. Hence the scattering remains finite. Without screening by the atomic electrons, the range of the Coulomb potential is infinite and $\sigma(\theta)$ would diverge as $\theta \to 0$ $(q \to 0)$.

Problem 13.3 If alpha particles are scattered from a lithium nucleus of radius 10^{-12} cm, what momentum exchange will occur at 30° if the initial alpha particle energy is 5.3 Mev in the center of mass system? In the laboratory system?

Problem 13.4 Calculate alpha particle scattering by a lead atom. The atomic potential may be approximated as $(zZe^2/r)e^{-r/r_0}$ where $r_0 \sim 1$ Å.

Problem 13.5 Use the Born approximation to calculate neutron–proton scattering if their interaction potential is $V_0(e^{-(\hbar/mc)r})/r$ where m is the mass of the π meson (273 electron masses). Calculate the scattering resulting if the potential were $V_0e^{-\hbar r/mc}$.

13.7 SCATTERING BY IDENTICAL PARTICLES

If a particle is scattered into angle θ by an identical target particle, which simultaneously scatters into angle $\pi - \theta$, the event cannot be distinguished from the similar event in which the target particle emerges at angle θ (see Figure 13.5). Therefore, one might think that the total expected differential scattering would be not $\sigma(\theta) = |f(\theta)|^2$ but (neglecting spin interactions)

$$\sigma(\theta)\,d\omega = |f(\theta) + f(\pi - \theta)|^2\,d\omega \qquad (13.37)$$

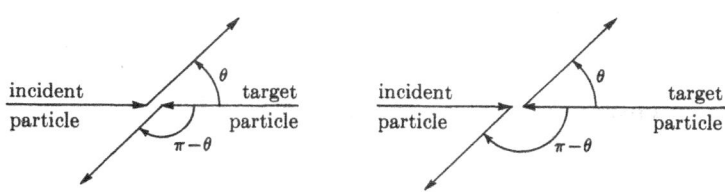

Figure 13.5 *Illustration of two scattering events indistinguishable in the center of mass coordinate system.*

This result is incomplete, however, if the two particles each have odd half-integral spin, inasmuch as then they must be in antisymmetric states. There are two possibilities: Either the particles are in antisymmetric spin states and Eq. (13.37), which represents scattering in a symmetric space state, is correct; or the particles are in a symmetric spin state and the scattering is given by the antisymmetric space function:

$$\sigma(\theta)\,d\omega = |f(\theta) - f(\pi - \theta)|^2\,d\omega \qquad (13.38)$$

Similar considerations apply to particles of integer spin which must be in a combined spin-space symmetrical state. The wave amplitudes for the two situations of (13.37, 13.38) cannot both be present to interfere with one another; for since the spin states are quantized, the complete wave function for the two particles in any individual scattering event must be either completely symmetrical or antisymmetrical in spin. Hence, the total space state must be correspondingly antisymmetric or symmetric, and interference cannot occur between states of different space symmetry. (If the particle spin symmetry is changed in the interaction, however, the space symmetry of the final states may differ from the space symmetry of the initial states. All the foregoing is valid only if there is no spin interaction.)

If electrons whose spins are random are scattered, there will be three times as many scatterings in the symmetrical spin states $\alpha\alpha$, $\beta\beta$, and $(\alpha\beta + \alpha\beta)$ as in the antisymmetric state $(\alpha\beta - \beta\alpha)$. Therefore, assuming no electron spin interaction, the total scattering will be given by

$$\sigma(\theta)\, d\omega = \{\tfrac{3}{4}|f(\theta) - f(\pi - \theta)|^2 + \tfrac{1}{4}|f(\theta) + f(\pi - \theta)|^2\}\, d\omega \quad (13.39)$$

Mott has derived electron–electron scattering exactly and obtained an additional factor in the interference term:

$$\sigma(\theta)\, d\omega = \left(\frac{e^2}{mv^2}\right)^2 \left\{ \frac{1}{\sin^4 \theta/2} + \frac{1}{\cos^4 \theta/2} - \frac{\cos\left[\dfrac{2e^2}{\hbar v}\log \tan \theta/2\right]}{\sin^2 \theta/2 \cos^2 \theta/2} \right\} d\omega \quad (13.40)$$

13.8 THE METHOD OF PARTIAL WAVES

Though convenient and simple to apply, the Born approximation is invalid for many problems of physical interest. Relativistic effects invalidate the assumption of static potentials at high energies; and at low energies it is incorrect to treat the potential energy of interaction as a small perturbation on the total energy of the interacting particles. The exact general treatment of the collision problem at low and medium energies where the static potential applies is based on phase shifts of the radial wave function. Since we are concerned only with central (angle-independent) potentials in this introductory account of scattering theory, we will be concerned primarily with the radial wave function and its wave equation. At large distances from the scattering center the radial wave equation is sinusoidal independently of whether or not a scattering potential exists. If a scattering center is present, however, the phase of the argument of the sinusoidal function is shifted by a constant amount over that for a free particle.

In a central potential where $V(\mathbf{r}) = V(r)$, the wave equation is easily transformed by the substitution

$$\psi_l(\mathbf{r}) = \frac{u_l(r)}{r} Y_l{}^m(\theta, \varphi) \qquad (13.41)$$

to

$$\frac{d^2 u_l}{dr^2} + \frac{2m}{\hbar^2}[E - V(r)]u_l - \frac{l(l+1)}{r^2} u_l = 0 \qquad (13.42)$$

where m is, as usual, the reduced particle mass $m_1 m_2/(m_1 + m_2)$, and the $Y_l{}^m$ are, as before, the spherical harmonic wave functions with quantum numbers l and m. The solutions to Eq. (13.42) are subject to the usual boundary conditions: the ψ's are finite at the origin and zero at infinity so that

$$u(0) = 0, \qquad u(\infty) = 0 \qquad (13.43)$$

The effects of spin and symmetry on the scattering, discussed in the preceding section, may be included by writing the free particle wave functions for odd or even l, whichever is appropriate. That is, in (13.42) one could make $V = 0$ for odd or even l, whichever is appropriate. Since the spatial state is specified by $(-1)^l$ as antisymmetric for $(-1)^l = -1$ and symmetric for $(-1)^l = +1$, protons in symmetric spin states, for example, would experience no interaction with one another in even l states.

The expected asymptotic behavior of ψ far beyond the scattering center (see Section 13.2) is a sum of a plane wave and a spherically outgoing wave,

$$\psi(r, \theta, \varphi) \to A\left[e^{ikz} + \frac{e^{ikr}}{r} f(\theta)\right] \qquad (13.44)$$

Hence, we expect that at large r, $u_l(r)$ will assume the form

$$u_l(r) = v_l(r)e^{\pm ikr} \qquad (13.45)$$

where $k = p/\hbar = \sqrt{2mE/\hbar^2}$, and $v_l(r)$ is a slowly varying function of r as we will now show. Substituting (13.45) into (13.42) we obtain

$$v_l'' \pm 2ikv_l' = \left[\frac{2m}{\hbar^2}V(r) + \frac{l(l+1)}{r^2}\right]v_l \qquad (13.46)$$

If v is slowly varying, v'' may be neglected and the rest of Eq. (13.46) may be solved by quadrature:

$$\pm 2ik \log v = \int^r \left[\frac{2m}{\hbar^2} V(r) + \frac{l(l+1)}{r^2} \right] dr \tag{13.47}$$

Since $V(r)$ approaches zero as r goes to infinity, v_l does indeed approach a constant for large r, thus justifying our assumption that v_l is slowly varying at large r. Therefore, the solutions to Eq. (13.42) at large particle separations become

$$u_l(r) = A_l' \sin (kr + \delta_l') \tag{13.48}$$

where δ_l' is a constant phase factor in the solution—one of the two arbitrary constants appearing in the solutions of all second-order differential equations. Since these constants are determined by the boundary conditions, we might expect, as is in fact the case, that δ_l' depends on the scattering potential; furthermore, the scattering cross-section may be put in terms of δ_l'. A direct connection between the scattering cross-section and the scattering potential does not exist in the exact theory. *The connection may be made only through the phase shift of the radial wave function, δ_l.*

The complete wave function may be expanded in terms of the orthonormal set of radial and angular wave functions,

$$\psi = \sum_{l=0}^{\infty} \frac{u_l(r)}{r} P_l(\cos \theta) \tag{13.49}$$

where $P_l(\cos \theta)$ is the Legendre polynomial of order l.

A slight modification of the theory is necessary in problems where $V(r)$ decreases with r faster than does $l(l+1)/r^2$. In this case the solution in the intermediate region where $V(r)$ is zero but $l(l+1)/r^2$ is appreciable is the solution to

$$\frac{d^2 u_l}{dr^2} + k^2 u_l - \frac{l(l+1)}{r^2} u_l = 0 \tag{13.50}$$

namely,

$$u_l(r) = \sqrt{\frac{\pi r}{2k}} A_l [(\cos \delta_l) J_{l+\frac{1}{2}}(kr) - (\sin \delta_l) N_{l+\frac{1}{2}}(kr)] \tag{13.51}$$

which is the radial solution to the spherical square well potential discussed in Chapter 8. $A_l \cos \delta_l$ and $A_l \sin \delta_l$ are two undetermined constants which are part of the solution to any second-order differential equation; δ_l must be fixed by the boundary conditions. For large r, the asymptotic form of the spheroidal Bessel and Neumann functions may be used, so that $u_l(r)$ becomes

$$u_l(r) = k^{-1} A_l \sin\left(kr - \frac{l\pi}{2} + \delta_l\right) \tag{13.52}$$

which again is the free particle wave function just as Eq. (13.48) is, with the difference that here the phase of the wave has been shifted by $[-(l\pi/2) + \delta_l]$ instead of by the phase shift computed by neglecting the centrifugal potential, δ_l'. This solution (Eqs. 13.48, 13.52) includes both the incident plane wave and the scattered spherical wave. The complete solution requires that one take the sum of $u_l(r)$ over all l. The scattering cross-section is found from the spherical wave contribution. To evaluate the scattering cross-section brought about by the scattering potential we must compute how much of (13.52) is due to the scattering potential. That is, we are interested only in the difference between the sum over l of (13.41) with (13.52) and the free particle wave functions. To find this difference we must first express the free particle plane wave function e^{ikz} in terms of functions similar to (13.41) and (13.52).

The plane wave e^{ikz} may be expanded in terms of the complete orthonormal set of radial Bessel functions of order $\frac{1}{2}$ and the Legendre polynomials. The expansion is

$$e^{ikz} = e^{ikr \cos\theta} = \sum_{l=0}^{\infty} (2l + 1)i^l \sqrt{\frac{\pi}{2kr}} J_{l+\frac{1}{2}}(kr) P_l(\cos\theta) \tag{13.53}$$

The asymptotic form of the Bessel function may be used at large r so that

$$e^{ikz} = \sum_{l=0}^{\infty} (2l + 1)i^l \frac{1}{kr} \sin\left(kr - \frac{l\pi}{2}\right) P_l(\cos\theta) \tag{13.53a}$$

Therefore, by substituting Eqs. (13.52) and (13.53a) into Eq. (13.44), we obtain

$$\sum_{l=0}^{\infty} (2l + 1)i^l \frac{1}{kr} \sin\left(kr - \frac{l\pi}{2}\right) P_l(\cos\theta) + \frac{1}{r}f(\theta)e^{ikr}$$

$$= \sum_{l=0}^{\infty} A_l \frac{1}{kr} \sin\left(kr - \frac{l\pi}{2} + \delta_l\right) P_l)\cos\theta) \tag{13.54}$$

which is the desired relation between $f(\theta)$ and A_l, δ_l we sought. Because of the term in e^{ikr}, it is useful to write the sine functions in terms of e^{+ikr} and e^{-ikr} so that each term in Eq. (13.54) is multiplied by either e^{ikr} or e^{-ikr} ($\sin kr = (e^{ikr} - e^{-ikr})/2i$). The equation is true for all values of k and large r, so that the equation must be true independently for the coefficients of e^{+ikr} and e^{-ikr}. Equating the coefficients of e^{+ikr}, we obtain

$$2ikf(\theta) + \sum_{l=0}^{\infty} (2l + 1)i^l e^{-il\pi/2} P_l(\cos\theta) = \sum_{l=0}^{\infty} A_l e^{i-i\delta_l(il\pi/2)} P_l(\cos\theta) \tag{13.55}$$

and equating the coefficients of e^{-ikr} we obtain

$$\sum_{l=0}^{\infty} (2l + 1)i^l e^{il\pi/2} P_l(\cos\theta) = \sum_{l=0}^{\infty} A_l e^{-i\delta_l+(il\pi/2)} P_l(\cos\theta) \tag{13.56}$$

Multiplying both sides of Eq. (13.56) by the particular Legendre polynomial $P_n(\cos\theta)$ and integrating over all angles, we find that because of the orthogonality of the P_l's,

$$A_n = (2n + 1)i^n e^{i\delta_n} \tag{13.57}$$

Substituting this result into Eq. (13.55), we may rewrite Eqs. (13.7) and (13.9) as

$$f(\theta) = \frac{1}{2ik} \sum_{l=0}^{\infty} \{-(2l + 1)i^l e^{-il\pi/2} P_l(\cos\theta) + (2l + 1)i^l e^{i\delta_l} e^{i\delta_l-(il\pi/2)} P_l(\cos\theta)\}$$

$$= \frac{1}{2ik} \sum_{l=0}^{\infty} (2l + 1)(e^{2i\delta_l} - 1)P_l(\cos\theta) \tag{13.58}$$

and

$$\sigma(\theta) = |f(\theta)|^2 = \frac{1}{k^2} \left| \sum_{l=0}^{\infty} (2l + 1)e^{i\delta_l}(\sin \delta_l)P_l(\cos \theta) \right|^2 \qquad (13.59)$$

which is the desired connection we have sought between the phase shift and the scattering cross-section. Equation (13.59) is Lord Rayleigh's formula for wave diffraction, which was rederived for quantum mechanics by H. Faxen and J. Holtzmark. Note that waves of different orbital momenta interfere with one another. This is an exact result for $\sigma(\theta)$; thus the scattering problem is reduced to finding the appropriate phase shift for each angular momentum. δ_l is the phase shift in the radial wave function (13.52) for a wave in a potential relative to a wave in no potential (it is *not* a space angle related to some direction in the scattering process!). δ_l is just a way of expressing the constants in the general solution which must be fitted to particular boundary conditions.

The total scattering cross-section is obtained by integrating Eq. (13.59) over all angles (note that the interference terms integrate to zero because of the orthogonality of the P_l ($\cos \theta$)):

$$\sigma = 2\pi \int_0^{\pi} \sigma(\theta) \sin \theta \, d\theta = \frac{4\pi}{k^2} \sum_{l=0}^{\infty} (2l + 1) \sin^2 \delta_l$$

$$= 4\pi\lambda^2 \sum_{l=0}^{\infty} (2l + 1) \sin^2 \delta_l \qquad (13.60)$$

In order to calculate the scattering cross-section, the solution to the wave equation within the range of the scattering potential, u_l, must be calculated and then fitted to the free-particle wave function (13.48) or (13.52) by an appropriate choice of δ_l. For example, if necessary, u_l could be integrated numerically for $r < a$, where a is the range of the potential, and then at $r = a$, u_l and its first derivative could be matched to Eq. (13.52) and its derivative, thus furnishing the boundary conditions needed to fix δ_l.

The radial wave function $u_l(r)$ at small r is an exponential function decreasing as $r \to 0$ for repulsive potentials, including the centripetal potential for which $u_l(r)$ behaves as r^{l+1} near $r = 0$; at large r, $u_l(r)$ behaves as $\sin [kr - (l\pi/2) + \delta_l]$. The various functions for a typical problem are illustrated in Figure 13.6 for $l = 0$. At very low energies, for which $k \to 0$ and $\lambda \to \infty$, the intercept of the tangent to the slope of the wave function at the edge of the potential $(r = a)$, b, illustrated in Figure 13.6, is approxi-

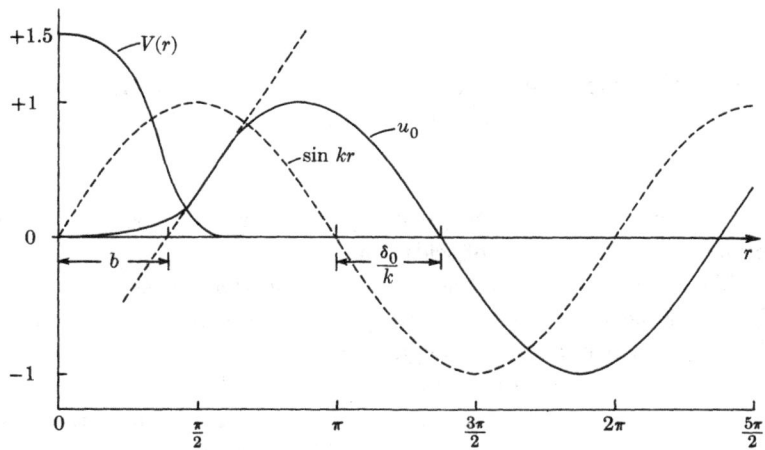

Figure 13.6   *The wave functions of a free particle and the wave function $u_l(r)$ of a particle influenced by a repulsive potential $V(r)$. $V(r)$ and the radial function u_l for $l = 0$ are shown in solid lines. The dashed sine curve is for the undisturbed free particle wave function, $\sin kr$. The problem illustrated here has a negative phase shift $-\delta_l$, although conceivably for other potentials or higher energies (and therefore short free-particle wavelengths compared to the range of the potential well) the phase shift could just as well be positive.*

mately equal to δ_0/k and is a constant independent of E or k since $\delta_0 \sim k$. The terms for $l \neq 0$ in the summation over l in the cross-section (13.59) are negligible and do not contribute to the sum for $k \ll a^{-1}$, since the penetration through the angular momentum barrier $\hbar^2 l(l + 1)/r^2$ is small, as will be discussed in the following sections. For this very important low-energy case, only δ_0 need be considered, that is, $\delta_0 = kb$ and approaches zero with k. Thus,

$$\sigma \xrightarrow[k \to 0]{\lim} \frac{4\pi}{k^2} k^2 b^2 = 4\pi b^2 \tag{13.61}$$

b has the dimensions of a length and is called the Fermi scattering length.

13.9 SCATTERING BY PARTICLE WAVES OF HIGH ORBITAL ANGULAR MOMENTUM

For large angular momentum quantum numbers we would expect a classical result. For this reason and for greater physical insight it is worthwhile to examine the classical theory of scattering.

Classically, the total scattering cross-section of a small incident particle would be πa^2 where a is the range of the interaction between incident particle and target object. The angular momentum of the incident particle with respect to the center of coordinates located in the target particle would be $\mathbf{r} \times \mathbf{p}$. If angular momentum were to be conserved, no particle with linear momentum p could be deflected which had an angular momentum $l\hbar$ greater than ap. That is, classically, for deflection in the potential field of the target particle we must have $l\hbar \leqslant ap$ ($l < a/\lambda$). Similarly, in quantum mechanics the particle wave functions for the radial wave equation for large l,

$$\frac{\partial^2 u_l}{\partial r^2} + \frac{2m}{\hbar^2}[E - V(r)]u_l - \frac{l(l + 1)}{r^2} u_l = 0 \tag{13.62}$$

tail off exponentially at small r and behave as r^{l+1} for very small r, and hence, although there may be some penetration into the potential region and therefore scattering of particles with large l, it is very small. That is, if $V(r) \ll E$ for $r > a$, then $\delta_l \ll 1$ for all $l \gg a/\lambda$. Nevertheless, although the scattering for $l \gg a/\lambda$ is small, it is of interest and the foregoing suggests a simple perturbation theory calculation for it.

The radial wave equation for a free particle in spherical coordinates is Eq. (13.50) and that for a particle in a potential is Eq. (13.42). Writing

v_l for u_l as the free particle wave function (see Eq. 13.50), we have

$$\frac{d^2 v_l}{dr^2} + \frac{2m}{\hbar^2} E v_l - \frac{l(l+1)}{r^2} v_l = 0 \tag{13.63}$$

$$\frac{d^2 u_l}{dr^2} + \frac{2m}{\hbar^2} E u_l - \frac{2m}{\hbar^2} V(r) u_l - \frac{l(l+1)}{r^2} u_l = 0 \tag{13.64}$$

Multiplying Eq. (13.63) by u_l and Eq. (13.64) by $-v_l$ and adding the results, we obtain

$$u_l \frac{d^2 v_l}{dr^2} - v_l \frac{d^2 u_l}{dr^2} + \frac{2m}{\hbar^2} V(r) v_l u_l = 0 \tag{13.65}$$

Integrating this equation over r we obtain

$$\left[u_l \frac{dv_l}{dr} - v_l \frac{du_l}{dr} \right]_0^R + \frac{2m}{\hbar^2} \int_0^R V(r) u_l v_l \, dr = 0 \tag{13.65a}$$

If the upper limit $R \gg a$, where R is taken arbitrarily far out from a, then the left-hand integrated term is independent of R, as we shall see. Evaluation at the lower limit gives 0, inasmuch as u_l and v_l are both zero at $r = 0$. At the upper limit, $v_l = \sin [kR - (\pi l/2)]$ and $u_l = \sin [kR - (\pi l/2) + \delta_l]$. The integrated expression in (13.65a) therefore becomes

$$\sin\left(kR - \frac{\pi l}{2} + \delta_l \right) k \cos\left(kR - \frac{\pi l}{2} \right) - \sin\left(kR - \frac{\pi l}{2} \right) k \cos\left(kR - \frac{\pi l}{2} + \delta_l \right)$$

$$= k \sin\left[\left(kR - \frac{\pi l}{2} + \delta_l \right) - \left(kR - \frac{\pi l}{2} \right) \right]$$

$$= k \sin \delta_l \tag{13.66}$$

Therefore, substituting in (13.65a),

$$k \sin \delta_l = -\frac{2m}{\hbar^2} \int_0^a u_l v_l V(r) \, dr \tag{13.67}$$

If δ_l is very small so that $v_l \sim u_l$, then the unknown δ_l in u_l may be neglected, so that Eq. (13.67) becomes

$$k \sin \delta_l = -\frac{2m}{\hbar^2} \int_0^a v_l^2 V(r) \, dr \tag{13.68}$$

which is just the result that would be obtained from perturbation theory. If $a \gg k^{-1}$ and $V(r)$ is slowly varying, the average value of v_l^2, $\frac{1}{2}$, may be used in Eq. (13.68). At large r,

$$\int v_l^2 V(r)\, dr \sim \tfrac{1}{2} \int V\, dr$$

and unless $V <$ (const./r) at large r, the integral will diverge. Hence, this method will not work for the Coulomb potential, which has been solved exactly by other means.

For particle energies sufficiently high that several l would contribute classically to the scattering cross-section (i.e., $ap/l \gg \hbar$, and the potential range is several wavelengths in radius), the total scattering cross-section assumes a particularly simple form. In the classical limit, δ_l is assumed large for $l < a/\lambda$ and negligible for $l > a/\lambda$. Thus the total cross-section, Eq. (13.60), becomes

$$\sigma = 4\pi\lambda^2 \sum_{l=0}^{a/\lambda} (2l + 1) \sin^2 \delta_l$$

$$\simeq 4\pi\lambda^2 \sum_{l=0}^{a/\lambda} (2l + 1)\tfrac{1}{2}$$

$$= 2\pi a^2 \tag{13.60a}$$

The geometrical cross-section is πa^2. Where does the extra πa^2 come from? It is the shadow scattering—the interference of the absorbed wave with the transmitted wave—and it occurs in optics in diffraction phenomena also. Thus πa^2 from the re-emitted wave plus another πa^2 due to shadow scattering add up to $2\pi a^2$.

Problem 13.6   Calculate the phase shifts δ_0 for a square well with the following conditions:

(a) ka arbitrary and V_0 very small but positive (that is, an attractive potential).

(b) ka arbitrary and V_0 very small but repulsive.

(c) What is the precise expression for the condition on V_0 which justifies the approximations used in (a) and (b)? What is V_0 less than?

(d) How would you solve this problem if V_0 were -10 Mev (repulsive!)? If E were 100 kev and a were 10^{-12} cm? For a neutron being scattered from an infinitely heavy nucleus? For the latter two conditions include $l > 0$ and obtain $\sigma(\theta)$ within $\sim 5\%$.

Problem 13.7 Calculate the scattering cross-section for low energy particles ($E = 0.200$ Mev) incident on a square well $V_0 = -20$ Mev, $a = 10^{-12}$ cm. How much does the $l = 1$ wave contribute to the differential cross-section if $E = 0.200$ Mev?

Problem 13.8 Calculate the scattering cross-section for low energy particles incident on a square well as a function of particle energy and the depth of the well. Assume that only s-wave ($l = 0$) scattering occurs at this energy. For what values of V_0 and E is the scattering a maximum, and why? If $\delta_0 = 180°$, what is the scattering cross-section? (This is essentially the Ramsauer effect discussed in Section 5.7.)

Problem 13.9 (a) Calculate approximately the contribution of the $l = 1$, 2, and 3 states to the scattering of 10 Mev neutrons by protons if their interaction potential is given by $V = -V_0 e^{-\hbar/mc}$, where m is the mass of the π meson (273 electron masses). Ignore any dependence of spin on nuclear forces. (b) Further suppose $\delta_0 = (\pi/2)$, $\delta_1 = (\pi/2)$, and $\delta_1 = 0$ for $l > 1$. What will be the fractional contributions of the two waves to the scattering? (c) How may δ_0 and δ_1 be separately determined experimentally?

Problem 13.10 Calculate the differential cross-section for low energy neutron–neutron scattering, assuming a square well attractive force. Neglect waves with $l > 2$.

Problem 13.11 Calculate the differential scattering cross-section for low energy alpha particles incident on helium nuclei, assuming a square well attractive force. Neglect waves with $l > 2$.

13.10 EFFECTIVE RANGE THEORY

By a modification of the previous section we may easily show that if the range of the interaction potential is very short compared to a wavelength, it is impossible to determine what the range or size of the potential is. This is an expected difficulty, since one can never determine spatial relationships to a precision better than the wavelength of the "radiation" one employs. By recognizing this difficulty, however, we can develop a general theory relating both a scattering cross-section or scattering length and an effective potential range for any reasonable potential shape, to the observed low energy experimental phase shifts δ_0. This effective range theory was first considered by Breit, but we will use the subsequent pellucid development by Bethe.

We wish to compare the values of the phase shifts for two different energy values E_1 and E_2 occurring in the wave equations

$$\frac{d^2u_1}{dr^2} + [k_1{}^2 + W(r)]u_1 = 0$$

$$\frac{d^2u_2}{dr^2} + [k_2{}^2 + W(r)]u_2 = 0 \qquad (13.69a)$$

where $k_1{}^2 = mE_1/\hbar^2$, $k_2{}^2 = mE_2/\hbar^2$, and $W(r) = -(m/\hbar^2)V(r)$. Multiplying the upper equation by u_2 and the lower by $-u_1$, and adding, we obtain

$$u_2\frac{d^2u_1}{dr^2} - u_1\frac{d^2u_2}{dr^2} + (k_1{}^2 - k_2{}^2)u_1u_2 = 0 \qquad (13.70)$$

Integrating over a dummy variable r' from 0 to r we obtain

$$\left[u_2\frac{du_1}{dr'} - u_1\frac{du_2}{dr'}\right]_0^r = (k_2{}^2 - k_1{}^2)\int_0^r u_1u_2\,dr' \qquad (13.71)$$

The lower limit of the integrated expression gives zero but the upper limit is nonzero. For large r,

$$u_1, u_2 \to \sin(k_{1,2}r + \delta_{1,2}) \qquad (13.72)$$

where in Eq. (13.72) the $\delta_{1,2}$ denote the phase shifts for the $l = 0$ wave function for states 1 and 2 respectively. For the region beyond the potential range, let ψ_1 and ψ_2 be solutions to the asymptotic differential equations.

$$\frac{d^2\psi_1}{dr^2} + k_1{}^2\psi_1 = 0 \quad \text{and} \quad \frac{d^2\psi_2}{dr^2} + k_2{}^2\psi_2 = 0 \qquad (13.73)$$

For convenience, let $\psi_1(0) = \psi_2(0) = 1$; then

$$\psi_1 = \frac{\sin(k_1 r + \delta_1)}{\sin \delta_1}, \qquad \psi_2 = \frac{\sin(k_2 r + \delta_2)}{\sin \delta_2} \qquad (13.74)$$

We want $\psi_{1,2}$ to be those solutions of Eq. (13.73) that agree with $u_{1,2}$ for larger r, that is, the solutions with the same phase shifts $\sin(k_{1,2}\, r + \delta_{1,2})$. ψ_1 and u_1 must be the same beyond the range of the potential; at $r = 0$, however, $\psi_1(0) \neq 0$, unlike u, by our condition and since anyway ψ_1 is not the correct solution to the wave equation near $r = 0$. Proceeding with Eqs. (13.73) as we did with Eqs. (13.64) in obtaining Eq. (13.71), we find

$$\left[\psi_2 \frac{d\psi_1}{dr'} - \psi_1 \frac{d\psi_2}{dr'} \right]_0^r = (k_2{}^2 - k_1{}^2) \int_0^r \psi_1 \psi_2 \, dr' \qquad (13.75)$$

We know that $u_1 = \psi_1$ and $u_2 = \psi_2$ at large r beyond the range of the potential. Therefore, the left-hand sides of (13.71) and (13.75) are equal for the upper limit r. The lower limit of the left-hand side of (13.71) gives zero. By subtracting Eq. (13.75) from Eq. (13.71) we obtain a result independent of the complicated behavior of the left-hand sides of these equations at large r,

$$- \left[\psi_2 \frac{d\psi_1}{dr'} - \psi_1 \frac{d\psi_2}{dr'} \right]_{r'=0} = k_2 \cot \delta_2 - k_1 \cot \delta_1$$

$$= (k_2{}^2 - k_1{}^2) \int_0^r (\psi_1 \psi_2 - u_1 u_2) \, dr' \qquad (13.76)$$

where $k_2 \cot \delta_2 - k_1 \cot \delta_1$ is obtained by substituting (13.74) in the left-hand side of (13.76).

Outside the range of nuclear forces this integral vanishes. If we define the effective range $\rho(E_1, E_2)$ by the equation

$$\rho(E_1, E_2) = 2 \int_0^\infty (\psi_1 \psi_2 - u_1 u_2) \, dr \qquad (13.77)$$

Eq. (13.76) becomes

$$k_2 \cot \delta_2 - k_1 \cot \delta_1 = \tfrac{1}{2}(k_2{}^2 - k_1{}^2)\rho(E_1, E_2) \qquad (13.76a)$$

a very useful relation. The δ's may be derived from the measured cross-sections; the effective range is a valuable theoretical quantity from which other cross-sections may be obtained and to which potentials may be related. Contributions to ρ all come from ψ and u at small values of r; the two wave functions are illustrated in Figure 13.7. For low energy at small r, ψ and u are essentially independent of the kinetic energy, so that we may use the zero energy wave functions in the effective range integral to obtain $\rho(0, 0) \equiv r_0$:

$$r_0 = 2 \int_0^\infty (\psi_0{}^2 - u_0{}^2) \, dr \qquad (13.78)$$

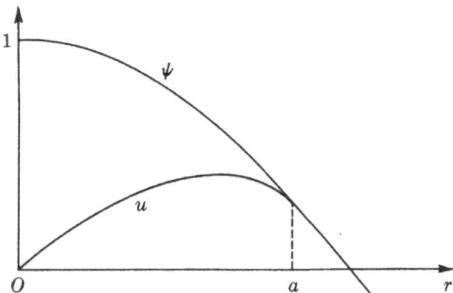

Figure 13.7 *Graph of ψ and u illustrating that they differ only near the origin where the potential is different from zero.*

At low energies ψ_0 approaches a straight line,

$$\psi_0 \xlongequal{r \sim 0} 1 + \left(\frac{\partial \psi}{\partial r} \right)_{r=0} r = 1 + (k \cot \delta)r$$

We define α to be

$$\alpha \xlongequal{\lim_{k \to 0}} k \cot \delta$$

α is the reciprocal of the Fermi scattering length,

$$b = \frac{1}{\alpha} \xlongequal{\lim_{\delta, k \to 0}} \frac{1}{k} \frac{\sin \delta}{\cos \delta}$$

If we then take $k_1 = \delta_1 = 0$ and drop the subscripts on k_2 and δ_2, Eq. (13.76a) yields

$$k \cot \delta = k_1 \cot \delta_1 + \tfrac{1}{2}k^2 r_0 = \alpha + \tfrac{1}{2}k^2 r_0 \qquad (13.79)$$

a relation which remains remarkably accurate up to 20 Mev, for example, for scattering of neutrons by protons. Thus the phase shifts can be calculated from a knowledge of the two parameters scattering length b and effective range r_0, which is the result we set out to obtain. A graph of u_0 and ψ_0 is presented in Figure 13.8.

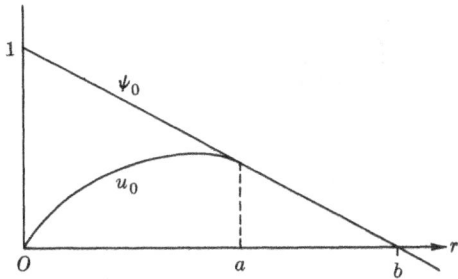

Figure 13.8   *Graph of ψ_0 and u_0 near the origin.*

An experimenter may interpret his scattering experiments by making a plot of $k \cot \delta$ as a function of k^2, which by Eq. (13.79) should yield the straight line illustrated in Figure 13.9. $k \cot \delta$ is obtained from the data by use of the relation between the observed scattering cross-section σ and the known particle energy k^2, Eq. (13.60), with $l = 0$:

$$\sigma = \frac{4\pi \sin^2 \delta}{k^2}$$

rearranging and noting that $\sin^{-2}\delta = 1 + \cot^2 \delta$ we obtain

$$\frac{4\pi}{\sigma} - k^2 = k^2 \cot^2 \delta$$

whence

$$k \cot \delta = \sqrt{\frac{4\pi}{\sigma} - k^2} \sim \sqrt{\frac{4\pi}{\sigma}} - \frac{1}{2}\sqrt{\frac{\sigma}{4\pi}}k^2 \qquad (13.80)$$

where $k \cot \delta$ has been put in terms of the measured quantities σ and k^2.

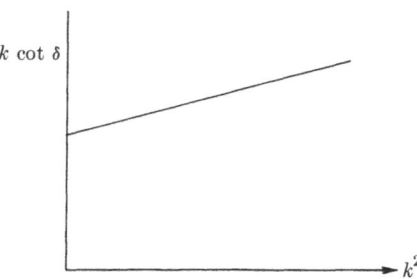

Figure 13.9   *Graph of $k \cot \delta$ vs. k^2.*

The intercept of the curve at $k^2 = 0$ in Figure 13.9 gives the reciprocal of the Fermi scattering length b. (From Eq. (13.60) with $l = 0$ we have

$$\sigma = \frac{4\pi}{k^2} \sin^2 \delta_0 \xrightarrow[k,\delta_0 \to 0]{\text{lim}} 4\pi b^2$$

in keeping with Eq. (13.61) so that b may be found from the experimentally determined cross-section σ, i.e., $b = \sqrt{\sigma/4\pi}$); the slope gives the effective range. The range and depth parameters of any possible theoretical potential well must be such that their wave functions give the measured effective range in Eq. (13.79), and the measured scattering length.

BIBLIOGRAPHY

Bohr, Niels, "The Penetration of Atomic Particles Through Matter," *Proc. Danish Royal Society*, **18**, No. 8 (1948).

Fermi, E., *Notes on Quantum Mechanics*. Chicago: University of Chicago Press, 1961.

Schiff, L. I., *Quantum Mechanics*. New York: McGraw-Hill Book Co., 1955.

Mott, N. F., and I. N. Snedden, *Wave Mechanics and Its Applications*. London: Oxford University Press, 1948.

Slater, J. C., *Quantum Theory of Atomic Structure*. New York: McGraw-Hill Book Co., 1960.

Pauling, L., and E. B. Wilson, *Introduction to Quantum Mechanics*. New York: McGraw-Hill Book Co., 1935.

APPENDIX:

THE VIBRATING STRING

AS AN ANALOGY to the more complicated three-dimensional problem of electromagnetic waves in a box we take up the problem of waves in a string. If a string fastened at both ends (at $x = 0$ and L) is displaced slightly in the y direction, the restoring force of an infinitesimal portion of the string, dx, will be given by the change in the y component of the tension on the string, F.

$$d\left[F\frac{dy}{\sqrt{(dx)^2 + (dy)^2}}\right] \sim d\left[F\frac{\partial y}{\partial x}\right] = F\frac{\partial^2 y}{\partial x^2}dx$$

The equation of motion for the infinitesimal string element is then given by

$$m\,dx\,\frac{\partial^2 y}{\partial t^2} = dF\,\frac{\partial y}{\partial x}$$

or

$$m\frac{\partial^2 y}{\partial t^2} = F\frac{\partial^2 y}{\partial x^2}$$

where m is the mass per unit length of the string.

We may attempt to solve this differential equation by assuming that y may be written as a product of a function of t times a function of x. The assumption is justified if a solution is obtained.

$$y = T(t)X(x); \quad m\,X(x)\frac{\partial^2 T(t)}{\partial t^2} = FT(t)\frac{\partial^2 X(x)}{\partial x^2}$$

which may be rewritten as

$$\frac{m}{F}\frac{1}{T(t)}\frac{\partial^2 T(t)}{\partial t^2} = \frac{1}{X(x)}\frac{\partial^2 X(x)}{\partial x^2}$$

The left-hand side of this equation is independent of x, the right-hand side is independent of t, and since they are equal, each must be independent of both x and t, and therefore constant. Let the constant be written as $-\omega^2$. Thus

$$\frac{m}{F}\frac{1}{T(t)}\frac{\partial^2 T(t)}{\partial t^2} = -\omega^2$$

or

$$\frac{\partial^2 T(t)}{\partial t^2} + \frac{F}{m}\omega^2 T(t) = 0$$

whose solution is

$$T(t) = A_1 \cos\sqrt{\frac{F}{m}}\omega t + A_2 \sin\sqrt{\frac{F}{m}}\omega t$$

where A_1 and A_2 are undetermined constants. We also have the equation for $X(x)$

$$\frac{\partial^2 X(x)}{\partial x^2} + \omega^2 X(x) = 0$$

whence

$$X(x) = B_1 \cos \omega x + B_2 \sin \omega x$$

where B_1 and B_2 are undetermined constants.

Since the string cannot vibrate at $x = 0$ and L, $X(0) = X(L) = 0$. Therefore, $B_1 = 0$ and ωL must be some integral multiple of π:

$$X(x) = B_{2n} \sin\frac{n\pi}{L}x$$

where n is an integer and there is a solution for every value of n.

The time dependence of the solution is determined by the initial conditions—the initial values of y and $\partial y/\partial t$ between $x = 0$ and $x = L$.

If we assume the string is vibrating at maximum amplitude at $t = 0$, then $A_2 = 0$, and we have ($y \neq 0$, $\partial y / \partial t = 0$)

$$T(t) = A_{1n} \cos \sqrt{\frac{F}{m} \frac{n\pi}{L}} t$$

Let $A_{1n}B_{2n} = C_n$; then the final complete solution may be written

$$y(x, t) = \sum_n C_n \cos \sqrt{\frac{F}{m} \frac{n\pi}{L}} t \sin \frac{n\pi}{L} x$$

The C_n depend on the shape in which the string is made to vibrate (whether it is plucked in the middle or near one end, etc.) and with what amplitude, and need not concern us for the present. What is important is that the frequencies (inverse of the period of the time dependence) are some discrete multiple of $\sqrt{(F/m)}(\pi/L)$; only these frequencies are permitted, as a result of the boundary conditions on a string of finite length.

It is also worthy of note that we have considered only one independent degree of freedom for the vibration of the string. An identical completely independent solution is obtained if we consider the vibration of the string in the z direction.

Mathematics-Bestsellers

HANDBOOK OF MATHEMATICAL FUNCTIONS: with Formulas, Graphs, and Mathematical Tables, Edited by Milton Abramowitz and Irene A. Stegun. A classic resource for working with special functions, standard trig, and exponential logarithmic definitions and extensions, it features 29 sets of tables, some to as high as 20 places. 1046pp. 8 x 10 1/2. 0-486-61272-4

ABSTRACT AND CONCRETE CATEGORIES: The Joy of Cats, Jiri Adamek, Horst Herrlich, and George E. Strecker. This up-to-date introductory treatment employs category theory to explore the theory of structures. Its unique approach stresses concrete categories and presents a systematic view of factorization structures. Numerous examples. 1990 edition, updated 2004. 528pp. 6 1/8 x 9 1/4. 0-486-46934-4

MATHEMATICS: Its Content, Methods and Meaning, A. D. Aleksandrov, A. N. Kolmogorov, and M. A. Lavrent'ev. Major survey offers comprehensive, coherent discussions of analytic geometry, algebra, differential equations, calculus of variations, functions of a complex variable, prime numbers, linear and non-Euclidean geometry, topology, functional analysis, more. 1963 edition. 1120pp. 5 3/8 x 8 1/2. 0-486-40916-3

INTRODUCTION TO VECTORS AND TENSORS: Second Edition--Two Volumes Bound as One, Ray M. Bowen and C.-C. Wang. Convenient single-volume compilation of two texts offers both introduction and in-depth survey. Geared toward engineering and science students rather than mathematicians, it focuses on physics and engineering applications. 1976 edition. 560pp. 6 1/2 x 9 1/4. 0-486-46914-X

AN INTRODUCTION TO ORTHOGONAL POLYNOMIALS, Theodore S. Chihara. Concise introduction covers general elementary theory, including the representation theorem and distribution functions, continued fractions and chain sequences, the recurrence formula, special functions, and some specific systems. 1978 edition. 272pp. 5 3/8 x 8 1/2. 0-486-47929-3

ADVANCED MATHEMATICS FOR ENGINEERS AND SCIENTISTS, Paul DuChateau. This primary text and supplemental reference focuses on linear algebra, calculus, and ordinary differential equations. Additional topics include partial differential equations and approximation methods. Includes solved problems. 1992 edition. 400pp. 7 1/2 x 9 1/4. 0-486-47930-7

PARTIAL DIFFERENTIAL EQUATIONS FOR SCIENTISTS AND ENGINEERS, Stanley J. Farlow. Practical text shows how to formulate and solve partial differential equations. Coverage of diffusion-type problems, hyperbolic-type problems, elliptic-type problems, numerical and approximate methods. Solution guide available upon request. 1982 edition. 414pp. 6 1/8 x 9 1/4. 0-486-67620-X

VARIATIONAL PRINCIPLES AND FREE-BOUNDARY PROBLEMS, Avner Friedman. Advanced graduate-level text examines variational methods in partial differential equations and illustrates their applications to free-boundary problems. Features detailed statements of standard theory of elliptic and parabolic operators. 1982 edition. 720pp. 6 1/8 x 9 1/4. 0-486-47853-X

LINEAR ANALYSIS AND REPRESENTATION THEORY, Steven A. Gaal. Unified treatment covers topics from the theory of operators and operator algebras on Hilbert spaces; integration and representation theory for topological groups; and the theory of Lie algebras, Lie groups, and transform groups. 1973 edition. 704pp. 6 1/8 x 9 1/4. 0-486-47851-3

Browse over 9,000 books at www.doverpublications.com

A SURVEY OF INDUSTRIAL MATHEMATICS, Charles R. MacCluer. Students learn how to solve problems they'll encounter in their professional lives with this concise single-volume treatment. It employs MATLAB and other strategies to explore typical industrial problems. 2000 edition. 384pp. 5 3/8 x 8 1/2.　　0-486-47702-9

NUMBER SYSTEMS AND THE FOUNDATIONS OF ANALYSIS, Elliott Mendelson. Geared toward undergraduate and beginning graduate students, this study explores natural numbers, integers, rational numbers, real numbers, and complex numbers. Numerous exercises and appendixes supplement the text. 1973 edition. 368pp. 5 3/8 x 8 1/2.　　0-486-45792-3

A FIRST LOOK AT NUMERICAL FUNCTIONAL ANALYSIS, W. W. Sawyer. Text by renowned educator shows how problems in numerical analysis lead to concepts of functional analysis. Topics include Banach and Hilbert spaces, contraction mappings, convergence, differentiation and integration, and Euclidean space. 1978 edition. 208pp. 5 3/8 x 8 1/2.　　0-486-47882-3

FRACTALS, CHAOS, POWER LAWS: Minutes from an Infinite Paradise, Manfred Schroeder. A fascinating exploration of the connections between chaos theory, physics, biology, and mathematics, this book abounds in award-winning computer graphics, optical illusions, and games that clarify memorable insights into self-similarity. 1992 edition. 448pp. 6 1/8 x 9 1/4.　　0-486-47204-3

SET THEORY AND THE CONTINUUM PROBLEM, Raymond M. Smullyan and Melvin Fitting. A lucid, elegant, and complete survey of set theory, this three-part treatment explores axiomatic set theory, the consistency of the continuum hypothesis, and forcing and independence results. 1996 edition. 336pp. 6 x 9.　　0-486-47484-4

DYNAMICAL SYSTEMS, Shlomo Sternberg. A pioneer in the field of dynamical systems discusses one-dimensional dynamics, differential equations, random walks, iterated function systems, symbolic dynamics, and Markov chains. Supplementary materials include PowerPoint slides and MATLAB exercises. 2010 edition. 272pp. 6 1/8 x 9 1/4.　　0-486-47705-3

ORDINARY DIFFERENTIAL EQUATIONS, Morris Tenenbaum and Harry Pollard. Skillfully organized introductory text examines origin of differential equations, then defines basic terms and outlines general solution of a differential equation. Explores integrating factors; dilution and accretion problems; Laplace Transforms; Newton's Interpolation Formulas, more. 818pp. 5 3/8 x 8 1/2.　　0-486-64940-7

MATROID THEORY, D. J. A. Welsh. Text by a noted expert describes standard examples and investigation results, using elementary proofs to develop basic matroid properties before advancing to a more sophisticated treatment. Includes numerous exercises. 1976 edition. 448pp. 5 3/8 x 8 1/2.　　0-486-47439-9

THE CONCEPT OF A RIEMANN SURFACE, Hermann Weyl. This classic on the general history of functions combines function theory and geometry, forming the basis of the modern approach to analysis, geometry, and topology. 1955 edition. 208pp. 5 3/8 x 8 1/2.　　0-486-47004-0

THE LAPLACE TRANSFORM, David Vernon Widder. This volume focuses on the Laplace and Stieltjes transforms, offering a highly theoretical treatment. Topics include fundamental formulas, the moment problem, monotonic functions, and Tauberian theorems. 1941 edition. 416pp. 5 3/8 x 8 1/2.　　0-486-47755-X

Browse over 9,000 books at www.doverpublications.com

Physics

THEORETICAL NUCLEAR PHYSICS, John M. Blatt and Victor F. Weisskopf. An uncommonly clear and cogent investigation and correlation of key aspects of theoretical nuclear physics by leading experts: the nucleus, nuclear forces, nuclear spectroscopy, two-, three- and four-body problems, nuclear reactions, beta-decay and nuclear shell structure. 896pp. 5 3/8 x 8 1/2.   0-486-66827-4

QUANTUM THEORY, David Bohm. This advanced undergraduate-level text presents the quantum theory in terms of qualitative and imaginative concepts, followed by specific applications worked out in mathematical detail. 655pp. 5 3/8 x 8 1/2.
0-486-65969-0

ATOMIC PHYSICS AND HUMAN KNOWLEDGE, Niels Bohr. Articles and speeches by the Nobel Prize–winning physicist, dating from 1934 to 1958, offer philosophical explorations of the relevance of atomic physics to many areas of human endeavor. 1961 edition. 112pp. 5 3/8 x 8 1/2.   0-486-47928-5

COSMOLOGY, Hermann Bondi. A co-developer of the steady-state theory explores his conception of the expanding universe. This historic book was among the first to present cosmology as a separate branch of physics. 1961 edition. 192pp. 5 3/8 x 8 1/2.
0-486-47483-6

LECTURES ON QUANTUM MECHANICS, Paul A. M. Dirac. Four concise, brilliant lectures on mathematical methods in quantum mechanics from Nobel Prize-winning quantum pioneer build on idea of visualizing quantum theory through the use of classical mechanics. 96pp. 5 3/8 x 8 1/2.   0-486-41713-1

THE PRINCIPLE OF RELATIVITY, Albert Einstein and Frances A. Davis. Eleven papers that forged the general and special theories of relativity include seven papers by Einstein, two by Lorentz, and one each by Minkowski and Weyl. 1923 edition. 240pp. 5 3/8 x 8 1/2.   0-486-60081-5

PHYSICS OF WAVES, William C. Elmore and Mark A. Heald. Ideal as a classroom text or for individual study, this unique one-volume overview of classical wave theory covers wave phenomena of acoustics, optics, electromagnetic radiations, and more. 477pp. 5 3/8 x 8 1/2.   0-486-64926-1

THERMODYNAMICS, Enrico Fermi. In this classic of modern science, the Nobel Laureate presents a clear treatment of systems, the First and Second Laws of Thermodynamics, entropy, thermodynamic potentials, and much more. Calculus required. 160pp. 5 3/8 x 8 1/2.   0-486-60361-X

QUANTUM THEORY OF MANY-PARTICLE SYSTEMS, Alexander L. Fetter and John Dirk Walecka. Self-contained treatment of nonrelativistic many-particle systems discusses both formalism and applications in terms of ground-state (zero-temperature) formalism, finite-temperature formalism, canonical transformations, and applications to physical systems. 1971 edition. 640pp. 5 3/8 x 8 1/2.   0-486-42827-3

QUANTUM MECHANICS AND PATH INTEGRALS: Emended Edition, Richard P. Feynman and Albert R. Hibbs. Emended by Daniel F. Styer. The Nobel Prize–winning physicist presents unique insights into his theory and its applications. Feynman starts with fundamentals and advances to the perturbation method, quantum electrodynamics, and statistical mechanics. 1965 edition, emended in 2005. 384pp. 6 1/8 x 9 1/4.   0-486-47722-3

Browse over 9,000 books at www.doverpublications.com

Physics

INTRODUCTION TO MODERN OPTICS, Grant R. Fowles. A complete basic undergraduate course in modern optics for students in physics, technology, and engineering. The first half deals with classical physical optics; the second, quantum nature of light. Solutions. 336pp. 5 3/8 x 8 1/2.   0-486-65957-7

THE QUANTUM THEORY OF RADIATION: Third Edition, W. Heitler. The first comprehensive treatment of quantum physics in any language, this classic introduction to basic theory remains highly recommended and widely used, both as a text and as a reference. 1954 edition. 464pp. 5 3/8 x 8 1/2.   0-486-64558-4

QUANTUM FIELD THEORY, Claude Itzykson and Jean-Bernard Zuber. This comprehensive text begins with the standard quantization of electrodynamics and perturbative renormalization, advancing to functional methods, relativistic bound states, broken symmetries, nonabelian gauge fields, and asymptotic behavior. 1980 edition. 752pp. 6 1/2 x 9 1/4.   0-486-44568-2

FOUNDATIONS OF POTENTIAL THERY, Oliver D. Kellogg. Introduction to fundamentals of potential functions covers the force of gravity, fields of force, potentials, harmonic functions, electric images and Green's function, sequences of harmonic functions, fundamental existence theorems, and much more. 400pp. 5 3/8 x 8 1/2.   0-486-60144-7

FUNDAMENTALS OF MATHEMATICAL PHYSICS, Edgar A. Kraut. Indispensable for students of modern physics, this text provides the necessary background in mathematics to study the concepts of electromagnetic theory and quantum mechanics. 1967 edition. 480pp. 6 1/2 x 9 1/4.   0-486-45809-1

GEOMETRY AND LIGHT: The Science of Invisibility, Ulf Leonhardt and Thomas Philbin. Suitable for advanced undergraduate and graduate students of engineering, physics, and mathematics and scientific researchers of all types, this is the first authoritative text on invisibility and the science behind it. More than 100 full-color illustrations, plus exercises with solutions. 2010 edition. 288pp. 7 x 9 1/4.   0-486-47693-6

QUANTUM MECHANICS: New Approaches to Selected Topics, Harry J. Lipkin. Acclaimed as "excellent" (*Nature*) and "very original and refreshing" (*Physics Today*), these studies examine the Mössbauer effect, many-body quantum mechanics, scattering theory, Feynman diagrams, and relativistic quantum mechanics. 1973 edition. 480pp. 5 3/8 x 8 1/2.   0-486-45893-8

THEORY OF HEAT, James Clerk Maxwell. This classic sets forth the fundamentals of thermodynamics and kinetic theory simply enough to be understood by beginners, yet with enough subtlety to appeal to more advanced readers, too. 352pp. 5 3/8 x 8 1/2.   0-486-41735-2

QUANTUM MECHANICS, Albert Messiah. Subjects include formalism and its interpretation, analysis of simple systems, symmetries and invariance, methods of approximation, elements of relativistic quantum mechanics, much more. "Strongly recommended." – *American Journal of Physics*. 1152pp. 5 3/8 x 8 1/2.   0-486-40924-4

RELATIVISTIC QUANTUM FIELDS, Charles Nash. This graduate-level text contains techniques for performing calculations in quantum field theory. It focuses chiefly on the dimensional method and the renormalization group methods. Additional topics include functional integration and differentiation. 1978 edition. 240pp. 5 3/8 x 8 1/2.   0-486-47752-5

Browse over 9,000 books at www.doverpublications.com

Physics

MATHEMATICAL TOOLS FOR PHYSICS, James Nearing. Encouraging students' development of intuition, this original work begins with a review of basic mathematics and advances to infinite series, complex algebra, differential equations, Fourier series, and more. 2010 edition. 496pp. 6 1/8 x 9 1/4.　　　　　　　　0-486-48212-X

TREATISE ON THERMODYNAMICS, Max Planck. Great classic, still one of the best introductions to thermodynamics. Fundamentals, first and second principles of thermodynamics, applications to special states of equilibrium, more. Numerous worked examples. 1917 edition. 297pp. 5 3/8 x 8.　　　　　　　　0-486-66371-X

AN INTRODUCTION TO RELATIVISTIC QUANTUM FIELD THEORY, Silvan S. Schweber. Complete, systematic, and self-contained, this text introduces modern quantum field theory. "Combines thorough knowledge with a high degree of didactic ability and a delightful style." – *Mathematical Reviews.* 1961 edition. 928pp. 5 3/8 x 8 1/2.　　　　　　　　0-486-44228-4

THE ELECTROMAGNETIC FIELD, Albert Shadowitz. Comprehensive undergraduate text covers basics of electric and magnetic fields, building up to electromagnetic theory. Related topics include relativity theory. Over 900 problems, some with solutions. 1975 edition. 768pp. 5 5/8 x 8 1/4.　　　　　　　　0-486-65660-8

THE PRINCIPLES OF STATISTICAL MECHANICS, Richard C. Tolman. Definitive treatise offers a concise exposition of classical statistical mechanics and a thorough elucidation of quantum statistical mechanics, plus applications of statistical mechanics to thermodynamic behavior. 1930 edition. 704pp. 5 5/8 x 8 1/4.
0-486-63896-0

INTRODUCTION TO THE PHYSICS OF FLUIDS AND SOLIDS, James S. Trefil. This interesting, informative survey by a well-known science author ranges from classical physics and geophysical topics, from the rings of Saturn and the rotation of the galaxy to underground nuclear tests. 1975 edition. 320pp. 5 3/8 x 8 1/2.
0-486-47437-2

STATISTICAL PHYSICS, Gregory H. Wannier. Classic text combines thermodynamics, statistical mechanics, and kinetic theory in one unified presentation. Topics include equilibrium statistics of special systems, kinetic theory, transport coefficients, and fluctuations. Problems with solutions. 1966 edition. 532pp. 5 3/8 x 8 1/2.
0-486-65401-X

SPACE, TIME, MATTER, Hermann Weyl. Excellent introduction probes deeply into Euclidean space, Riemann's space, Einstein's general relativity, gravitational waves and energy, and laws of conservation. "A classic of physics." – *British Journal for Philosophy and Science.* 330pp. 5 3/8 x 8 1/2.　　　　　　　　0-486-60267-2

RANDOM VIBRATIONS: Theory and Practice, Paul H. Wirsching, Thomas L. Paez and Keith Ortiz. Comprehensive text and reference covers topics in probability, statistics, and random processes, plus methods for analyzing and controlling random vibrations. Suitable for graduate students and mechanical, structural, and aerospace engineers. 1995 edition. 464pp. 5 3/8 x 8 1/2.　　　　　　　　0-486-45015-5

PHYSICS OF SHOCK WAVES AND HIGH-TEMPERATURE HYDRO DYNAMIC PHENOMENA, Ya B. Zel'dovich and Yu P. Raizer. Physical, chemical processes in gases at high temperatures are focus of outstanding text, which combines material from gas dynamics, shock-wave theory, thermodynamics and statistical physics, other fields. 284 illustrations. 1966–1967 edition. 944pp. 6 1/8 x 9 1/4.
0-486-42002-7

Browse over 9,000 books at www.doverpublications.com